THE CARIBOO TRAIL

THE CARIBOO TRAIL

A CHRONICLE OF THE GOLD-FIELDS
OF BRITISH COLUMBIA

AGNES C. LAUT

Foreword by
DIANA FRENCH

TouchWood
Editions

TouchWood Editions
touchwoodeditions.com

LIBRARY AND ARCHIVES CANADA CATALOGUING IN PUBLICATION
Laut, Agnes C. (Agnes Christina), 1871–1936
The Cariboo trail: a chronicle of the gold-fields of
British Columbia / Agnes C. Laut.

Originally published: Toronto: Glasgow, Brook, 1916.
(Chronicles of Canada; 23)
Includes bibliographical references and index.
Issued also in electronic formats.
ISBN 978-1-77151-033-2

1. Cariboo (BC: Regional district)—Gold discoveries.
2. Gold mines and mining—British Columbia—Cariboo
(Regional district)—History. 3. Frontier and pioneer life—
British Columbia—Cariboo (Regional district). I. Title.

FC3822.5.L39 2013   971.1'7502   C2013-902128-0

Proofreader: Holland Gidney

Canadian Heritage   Patrimoine canadien   Canada Council for the Arts   Conseil des Arts du Canada   BRITISH COLUMBIA ARTS COUNCIL

We gratefully acknowledge the financial support for our publishing activities from the Government of Canada through the Canada Book Fund, Canada Council for the Arts, and the province of British Columbia through the British Columbia Arts Council and the Book Publishing Tax Credit.

MIX
Paper from responsible sources
FSC   FSC® C016245
www.fsc.org

The interior pages of this book have been printed on 30% post-consumer recycled paper, processed chlorine free, and printed with vegetable-based inks.

1 2 3 4 5 17 16 15 14 13

PRINTED IN CANADA

The first Legislative Assembly of Vancouver Island
Back Row—J.W. McKay, J.D. Pemberton, J. Porter (Clerk)
Front Row—T.J. Skinner, J.S. Helmcken, MD, James Yates
AFTER A PHOTOGRAPH

CONTENTS

�҈   �҈   ✧

ILLUSTRATIONS

Map of the Cariboo Country

MAP BY BARTHOLOMEW

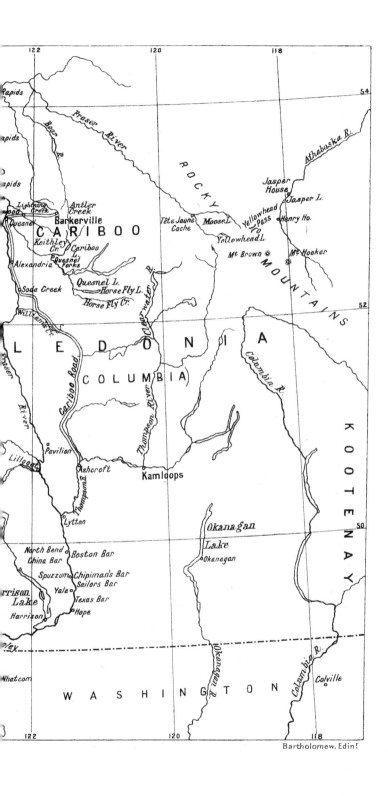
Bartholomew, Edin.

FOREWORD
by Diana French

Many tales have been told of the Cariboo Gold Rush. *The Cariboo Trail: A Chronicle of the Gold-fields of British Columbia* by Agnes C. Laut is one of the better ones. In it she paints vivid pictures of the successes and failures of the men who, bitten by the gold bug, battled the uncompromising landscapes, unfriendly Natives, and sometimes each other in their quest for the precious metal.

Laut (1871–1936) was an established freelance writer and the author of numerous novels and histories of the Canadian frontier. She wrote *The Cariboo Trail*, published in 1916, as part of the thirty-two volume Chronicles of Canada. She was born in Ontario, and raised in Winnipeg, Manitoba, then a frontier town, but because of her work she later made her home in New York State. Laut loved the outdoors and spent months each year on extended camping trips, many in the northwest. She travelled the pioneer trails of British Columbia where she met veterans of the gold rush. Their stories fascinated her.

Not all the stories are happy ones. Only one prospector in a thousand struck it rich and some of them died penniless. Many simply gave up and went home. It is not known how many perished from drowning, landslides, mishaps, or starvation. What is known is that the gold-rush miner "was an instrument in the hands of destiny, an instrument for shaping empire; for it was the inrush of miners which gave birth to the colony of British Columbia."

Laut's narrative begins in March 1858 in Victoria, population 500, headquarters of the Hudson's Bay fur-trading company, where James Douglas presided as Chief Factor and Governor of what is now British Columbia. A few prospectors had drifted north from the California goldfields by this time, but no one paid them much attention until significant amounts of gold dust and nuggets started appearing at the San Francisco mint. Then a "mad rabble of gold-crazy prospectors" arrived in Victoria hell-bent on reaching the Fraser River.

Thousands followed. They travelled from Victoria to Fort Langley by Hudson's Bay steamers, then upriver to Yale by canoe, rafts, and steamship. From there they scrambled up the treacherous Fraser River canyon as best they could, following the "float" (specks of gold in the sand) they hoped would lead them to the mother lode. Given the huge number of men involved and the competition for claims, there was surprisingly little lawlessness. Governor Douglas, the Royal Engineers, and later the infamous Judge Matthew Begbie were quick to ensure at least minimal law and order.

By the autumn of 1859, prospectors had reached Quesnel and Fort George and were exploring creeks, streams, and alpine estuaries. In the spring of 1861, a huge gold deposit found at Antler Creek "set the world agog." The Cariboo Gold Rush was on.

"The Cariboo became, in popular imagination, a land where "nuggets grew on the side of the road and could be picked up in baskets." The outside world heard romance, not hardships. Men whose "only knowledge of gold was that it was yellow" swarmed into the territory. By 1862, six thousand prospectors, most of them greenhorns, were in the Cariboo. They came from all over Canada, the USA, and Europe.

"How these people ever gained access to the centre of the wilderness before the famous Cariboo Road had been built" was a mystery to Laut. While most would-be miners reached the goldfields from the south, one party trekked across the country. In June 1862, one hundred and fifty Overlanders left Fort Garry, Manitoba, and headed for the Cariboo in wagons pulled by oxen. For them, "Far-off gold glittered the brighter for the distance." They spent four months crossing the prairie, battling floods, muskegs, mosquitoes, the Rocky Mountains, and finally the upper Fraser River. The survivors reached their destinations in Quesnel and Kamloops before winter, but none was known to strike it rich.

Getting supplies into the Cariboo and getting the gold out was a daunting task. In 1860, Governor Douglas had the Royal Engineers begin blasting the four-hundred-and-eighty-mile-long Cariboo Road through the formidable Fraser Canyon. Laut calls it one of the wonders of the world. It was completed in 1862, and the Cariboo was open for business.

Entrepreneurs found ways to profit from the gold, some legitimate, some not. There were many "claim jumping species of human vultures living off other men's risks." Life in the mining camps was rough but Judge Begbie, known for dispensing justice as much as law, kept the frontier more or less under control.

With the completion of the Cariboo Road, travel became safer, but the journey was still no joy, as is clear from Laut's description of stagecoach transportation:

> The horses sprang forward; and away the stage rattled round curves where a hind wheel would try to go over the edge—only the driver didn't let it; down embankments where any normal wagon would have upset, but this one didn't; up sharp grades where no horses ought to be driven at a trot, but where the six persisted in going at a gallop! The passenger didn't mind the jolting that almost dislocated his spine. He was thankful not to be held up by highwaymen, or dumped into the wild cataract of waters below.

Laut gives the women of the era their due. Be they miners' wives or dance-hall girls, they were relatively safe, she says, because under the "rough chivalry of an unwritten law" no man dared "wrong them."

Over the twelve years of the gold rush, the camps changed "from a poor man's camp to a camp for a capitalist or a company." Laut explains. "The miners first found the gold in flakes, then farther up in nuggets, then the nuggets had to be pursued to pay-dirt beneath gravel and clay." Shafts, tunnels, hydraulic machinery, and stamp mills replaced pick axes, gold pans, and sluice boxes. "Later, when the pay-dirt showed signs of merging into quartz, there passed away for ever the day of the penniless prospector seeking the golden fleece of the hills." When BC joined the Canadian Dominion in 1871, "the railway and the settler came; and the man with the pick and his eyes on the 'float' gave place to the man with the plough."

Laut was one of the few female historical writers a century ago. The gold rush was very much a "he" event but she sees it from a woman's perspective. Her dramatic imagery takes the reader along the trail with the prospectors, and while her colonial attitude can sometimes be a bit jarring by today's sensitivities, she was writing in colonial times, and that attitude too is part of BC history.

The Cariboo Trail is both educational and a good read because of the story Laut tells and the manner in which she tells it.

CHAPTER ONE
THE 'ARGONAUTS'

Early in 1849 the sleepy quiet of Victoria, Vancouver Island, was disturbed by the arrival of straggling groups of ragged nondescript wanderers, who were neither trappers nor settlers. They carried blanket packs on their backs and leather bags belted securely round the waist close to their pistols. They did not wear moccasins after the fashion of trappers, but heavy, knee-high, hobnailed boots. In place of guns over their shoulders, they had picks and hammers and such stout sticks as mountaineers use in climbing. They did not forgather with the Indians. They shunned the Indians and had little to say to any one. They volunteered little information as to whence they had come or whither they were going. They sought out Roderick Finlayson, chief trader for the Hudson's Bay Company. They wanted provisions from the company—yes—rice, flour, ham, salt, pepper, sugar, and tobacco; and at the smithy they demanded shovels, picks, iron ladles, and wire screens. It was only when they came to pay that Finlayson felt sure of what he had already guessed. They unstrapped those little leather bags round under their cartridge belts and produced in tiny gold nuggets the price of what they had bought.

Finlayson did not know exactly what to do. The fur-trader hated the miner. The miner, wherever he went, sounded the knell of fur-trading; and the trapper did not like to have his game preserve overrun by fellows who scared off all animals from traps, set fire going to clear away underbrush, and owned responsibility to no authority. No doubt these men were 'argonauts' drifted up from the gold diggings of California; no doubt they were searching for new mines; but who had ever heard of gold in Vancouver Island, or in New Caledonia, as the mainland was named? If there had been gold, would not the company have found it? Finlayson probably thought the easiest way to get rid of the unwelcome visitors was to let them go on into the dangers of the wilds and then spread the news of the disappointment bound to be theirs.

He handled their nuggets doubtfully. Who knew for a certainty that it was gold anyhow? They bade him lay it on the smith's anvil and strike it with a hammer. Finlayson, smiling sceptically, did as he was told. The nuggets flattened to a yellow leaf as fine and flexible as silk. Finlayson took the nuggets at eleven dollars an ounce and sent the gold down to San Francisco, very doubtful what the real value would prove. It proved sixteen dollars to the ounce.

For seven or eight years afterwards rumours kept floating in to the company's forts of finds of gold. Many of the company's servants drifted away to California in the wake of the 'Forty-Niners,' and the company found it hard to keep its trappers from deserting all up and down the Pacific Coast. The quest for gold had become a sort of yellow-fever madness. Men flung certainty to the winds and trekked recklessly to California, to Oregon, to the hinterland of the country round Colville and Okanagan. Yet nothing occurred to cause any excitement in Victoria. There was a short-lived flurry over the discovery in Queen Charlotte Islands of a nugget valued at six hundred dollars and a vein of gold-bearing quartz. But the nugget was an isolated freak; the quartz could not be worked at a profit; and the movement suddenly died out. There were, however, signs of what was to follow. The chief trader at the little fur-post of Yale reported that when he rinsed sand round in his camp frying-pan, fine flakes and scales of yellow could be seen at the bottom.[1] But gold in such minute particles would not satisfy the men who were hunting nuggets. It required treatment by quicksilver. Though Maclean, the chief factor at Kamloops, kept all the specks and flakes brought to his post as samples from 1852 to 1856, he had less than would fill a half-pint bottle. If a half-pint is counted as a half-pound and the gold at the company's price of eleven dollars an ounce, it will be seen why four years of such discoveries did not set Victoria on fire.

It has been so with every discovery of gold in the history of the world. The silent, shaggy, ragged first scouts of the gold stampede wander houseless for years from hill to hill, from gully to gully, up rivers, up stream beds, up dry watercourses, seeking the source of those yellow specks seen far down the mountains near the sea. Precipice, rapids, avalanche, winter storm, take their toll of dead. Corpses are washed down in the spring floods; or the thaw reveals a prospector's shack smashed by a snowslide under which lie two dead 'pardners.' Then, by and by, when everybody has forgotten about it, a shaggy man comes out of the wilds with a leather bag; the bag goes to the mint; and the world goes mad.

Victoria went to sleep again. When men drifted in to trade dust and nuggets for picks and flour, the fur-traders smiled, and rightly surmised that the California diggings were playing out.

Though Vancouver Island was nominally a crown colony, it was still, with New Caledonia, practically a fief of the Hudson's Bay Company. James Douglas was governor. He was assisted in the administration by a council of three, nominated by himself—John Tod, James Cooper, and Roderick Finlayson. In 1856 a colonial legislature was elected and met at Victoria in August for the first time.[2] But, in fact, the company owned the colony, and its will was supreme in the government. John Work was the company's chief factor at Victoria and Finlayson was chief trader.

Because California and Oregon had gone American, some small British warships lay at Esquimalt harbour. The little fort had expanded beyond the stockade. The governor's house was to the east of the stockade. A new church had been built, and the Rev. Edward Cridge, afterwards known as Bishop Cridge, was the rector. Two schools had been built. Inside the fort were perhaps forty-five employees. Inside and outside lived some eight hundred people. But grass grew in the roads. There was no noise but the church bell or the fort bell, or the flapping of a sail while a ship came to anchor. Three hundred acres about the fort were worked by the company as a farm, which gave employment to about two dozen workmen, and on which were perhaps a hundred cattle and a score of brood mares. The company also had a saw-mill. Buildings of huge, squared timbers flanked three sides of the inner stockades—the dining-hall, the cook-house, the bunk-house, the store, the trader's house. There were two bastions, and from each cannon pointed. Close to the wicket at the main entrance stood the postoffice. Only a fringe of settlement went beyond the company's farm. The fort was sound asleep, secure in an eternal certainty that the domain which it guarded would never be overrun by American settlers as California and Oregon had been. The little Admiralty cruisers which lay at Esquimalt were guarantee that New Caledonia should never be stampeded into a republic by an inrush of aliens. Then, as now, it was Victoria's boast that it was more English than England.

So passed Christmas of '57 with plum-pudding and a roasted ox and toasts to the crown and the company, though we cannot be quite sure that the company was not put before the crown in the souls of the fur-traders.

Then, in March 1858, just when Victoria felt most secure as the capital of a perpetual fur realm, something happened. A few Yankee prospectors had gone down on the Hudson's Bay steamer *Otter* to San Francisco in February with gold dust and nuggets from New Caledonia to exchange for money at the mint. The Hudson's Bay men had thought nothing of this. Other treasure-seekers had come to New Caledonia before and had gone back to San Francisco disappointed. But, in March, these men returned to Victoria. And with them came a mad rabble of gold-crazy prospectors. A city of tents sprang up overnight round Victoria. The smithy was besieged for picks, for shovels, for iron ladles. Men stood in long lines for their turn at the trading-store. By canoe, by dugout, by pack-horse, and on foot, they planned to ascend the Fraser, and they mobbed the company for passage to Langley by the first steamer out from Victoria. Goods were paid for in cash. Before Finlayson could believe his own eyes, he had two million dollars in his safe, some of it for purchases, some of it on deposit for safe keeping. Though the company gave no guarantee to the depositors and simply sealed each man's leather pouch as it was placed in the safe, no complaint was ever made against it of dishonesty or unfair treatment.

Without waiting instructions from England and with poignant memory of Oregon, Governor Douglas at once clapped on a licence of twenty-one shillings a month for mining privileges under the British crown. Thus he obtained a rough registration of the men going to the up-country; but thousands passed Victoria altogether and went in by pack-train from Okanagan or rafted across from Puget Sound. The month of March had not ended when the first band of gold hunters arrived and settled down a mile and a half below Yale. Another boat-load of eight hundred and fifty came in April. In four months sixty-seven vessels, carrying from a hundred to a thousand men each, had come up from San Francisco to Victoria. Crews deserted their ships, clerks deserted the company, trappers turned miners and took to the gold-bars. Before Victoria awoke to what it was all about, twenty thousand people were camped under tents outside the stockade, and the air was full of the wildest rumours of fabulous gold finds.

The snowfall had been heavy in '58. In the spring the Fraser rolled to the sea a swollen flood. Against the turbid current worked tipsy rafts towed by wheezy steamers or leaky old sailing craft, and rickety row-boats raced cockle-shell canoes for the gold-bars above. Ashore, the banks of the river were lined with foot passengers toiling under heavy packs, wagons to which clung human forms on every foot of space, and long rows of pack-horses bogged in the flood of the overflowing river. By September ten thousand men were rocking and washing for gold round Yale.

As in the late Kootenay and in the still later Klondike stampede, American cities at the coast benefited most. Victoria was a ten-hour trip from the mainland. Whatcom and Townsend, on the American side, advertised the advantages of the Washington route to the Fraser river gold-mines. A mushroom boom in town lots had sprung up at these points before Victoria was well awake. By the time speculators reached Victoria the best lots in that place had already been bought by the company's men; and some of the substantial fortunes of Victoria date from this period. Though the river was so high that the richest bars could not be worked till late in August, five hundred thousand dollars in gold was taken from the bed of the Fraser during the first six months of '58. This amount, divided among the ten thousand men who were on the bars around Yale, would not average as much as they could have earned as junior clerks with the fur company, or as peanut pedlars in San Francisco; but not so does the mind of the miner work. Here was gold to be scooped up for nothing by the first comer; and more vessels ploughed their way up the Fraser, though Governor Douglas sought to catch those who came by Puget Sound and evaded licence by charging six dollars toll each for all canoes on the Fraser and twelve dollars for each vessel with decks. Later these tolls were disallowed by the home authorities. The prompt action of Douglas, however, had the effect of keeping the mining movement in hand. Though the miners were of the same class as the 'argonauts' of California, they never broke into the lawlessness that compelled vigilance committees in San Francisco.

Judge Howay gives the letter of a treasure-seeker who reached the Fraser in April, the substance of which is as follows:

> We're now located thirty miles above the junction of the Fraser and the Thompson on Fraser River . . . About a fourth of the canoes that attempt to come up are lost in the rapids which extend from Fort Yale nearly to the Forks. A few days ago six men were drowned by their canoe upsetting. There is more danger going down than coming up. There can be no doubt about this country being immensely rich in gold. Almost every

Sir James Douglas
FROM A PORTRAIT BY SAVANNAH

bar on the river from Yale up will pay from three dollars to seven dollars a day to the man at the present stage of water. When the river gets low, which will be about August, the bars will pay very well. One hundred and ninety-six dollars was taken out by one man last winter in a few hours, but the water was then at its lowest stage. The gold on the bars is all very fine and hard to save in a rocker, but with quicksilver properly managed, good wages can be made almost anywhere on the river as long as the bars are actually covered with water. We have not yet been able to find a place where we can work anything but rockers. If we could get a sluice to work, we could make from twelve dollars to sixteen dollars a day each. We only commenced work yesterday and we are satisfied that when we get fully under way we can make from five dollars to seven dollars a day each. The prospect is better as we go up the river on the bars. The gold is not any coarser, but there is more of it. There are also in that region diggings of coarser gold on small streams that empty into the main river. A few men have been there and proved the existence of rich diggings by bringing specimens back with them. The Indians all along the river have gold in their possession that they say they dug themselves, but they will not tell where they get it, nor allow small parties to go up after it. I have seen pieces in their possession weighing two pounds. The Indians above are disposed to be troublesome and went into a camp twenty miles above us and forcibly took provisions and arms from a party of four men and cut two severely with their knives. They came to our camp the same day and insisted that we should trade with them or leave the country. We design to remain here until we can get a hundred men together, when we will move up above the falls and do just what we please without regard to the Indians. We are at present the highest up of any white men on the river, and we must go higher to be satisfied. I don't apprehend any danger from the Indians at present, but there will be hell to pay after a while. There is a pack-trail from Hope, but it cannot be travelled till the snow is off the mountains.

The prices of provisions are as follows: flour thirty-five

dollars per hundred-weight, pork a dollar a pound, beans fifty cents a pound, and other things in proportion. Every party that starts from the Sound should have their own supplies to last them three or four months, and they should bring the largest size chinook canoes, as small ones are very liable to swamp in the rapids. Each canoe should be provided with thirty fathoms of strong line for towing over swift water, and every man well armed. The Indians here can beat anything alive stealing. They will soon be able to steal a man's food after he has eaten it.

Within two miles of Yale eighty Indians and thirty white men were working the gold-bars; and log boarding-houses and saloons sprang up along the river-bank as if by magic. Naturally, the last comers of '58 were too late to get a place on the gold-bars, and they went back to the coast in disgust, calling the gold stampede 'the Fraser River humbug.' Nevertheless, men were washing, sluicing, rocking, and digging gold as far as Lillooet. Often the day's yield ran as high as eight hundred dollars a man; and the higher up the treasure-seekers pushed their way, the coarser grew the gold flakes and grains. Would the golden lure lead finally to the mother lode of all the yellow washings? That is the hope that draws the prospector from river to stream, from stream to dry gully bed, from dry gully to precipice edge, and often over the edge to death or fortune.

Exactly fifty-six years from the first rush of '58 in the month of April, I sat on the banks of the Fraser at Yale and punted across the rapids in a flat-bottomed boat and swirled in and out among the eddies of the famous bars. A Siwash family lived there by fishing with clumsy wicker baskets. Higher up could be seen some Chinamen, but whether they were fishing or washing we could not tell. Two transcontinental railroads skirted the canyon, one on each side, and the tents of a thousand construction workers stood where once were the camps of the gold-seekers banded together for protection. When we came back across the river an old, old man met us and sat talking to us on the bank. He had come to the Fraser in that first rush of '58.

Indians near New Westminster, BC
FROM A PHOTOGRAPH BY MAYNARD

He had been one of the leaders against the murderous bands of
Indians. Then, he had pushed on up the river to Cariboo, travelling,
as he told us, by the Indian trails over 'Jacob's ladders'—wicker and
pole swings to serve as bridges across chasms—wherever the 'float' or
sign of mineral might lead him. Both on the Fraser and in Cariboo
he had found his share of luck and ill luck; and he plainly regretted
the passing of that golden age of danger and adventure. 'But,' he
said, pointing his trembling old hands at the two railways, 'if we
prospectors hadn't blazed the trail of the canyon, you wouldn't have
your railroads here to-day. They only followed the trail we first cut
and then built. We followed the "float" up and they followed us.'

What the trapper was to the fur trade, the prospector was to the
mining era that ushered civilization into the wilds with a blare of
dance-halls and wine and wassail and greed. Ragged, poor, roofless,
grubstaked by 'pardner' or outfitter on a basis of half profit, the prospector stands as the eternal type of the trail-maker for finance.

CHAPTER TWO

THE PROSPECTOR

By September, when mountain rivers are at their lowest, every bar on the Fraser from Yale to the forks of the Thompson was occupied. The Hudson's Bay steamer *Otter* made regular trips up the Fraser to Fort Langley; and from the fort an American steamer called the *Enterprise*, owned by Captain Tom Wright, breasted the waters as far as the swift current at Yale. At Yale was a city of tents and hungry men. Walter Moberly tells how, when he ascended the Fraser with Wright in the autumn of '58, the generous Yankee captain was mobbed by penniless and destitute men for return passage to the coast. Many a broken treasure-seeker owed his life to Tom Wright's free passage. Fortunately, there was always good fishing on the Fraser; but salt was a dollar twenty-five a pound, butter a dollar twenty-five a pound, and flour rarer than nuggets. So hard up were some of the miners for pans to wash their gold, that one desperate fellow went to a log shack called a grocery store, and after paying a dollar for the privilege of using a grindstone, bought an empty butter vat at the pound price of butter—twelve dollars for an empty butter tub! Half a dollar was the smallest coin used, and clothing was so scarce that when a Chinaman's pig chewed up Walter Moberly's boots while the surveyor lay asleep in his shack, Mr Moberly had to foot it twenty-five miles before he could find another pair of boots. Saloons occupied every second shack at Yale and Hope; revolvers were in all belts and each man was his own sheriff; yet there was little lawlessness.

With claims filed on all gold-bearing bars, what were the ten thousand men to do camped for fifty miles beyond Yale? Those who had no provisions and could not induce any storekeeper to grubstake them for a winter's prospecting, quit the country in disgust; and the price of land dropped in the boom towns of the Fraser as swiftly as it had been ballooned up. Prospecting during the winter in a country of heavy snowfall did not seem a sane project. And yet the eternal question urged the miners on: from what mother lode are these flakes and nuggets washed down to the sand-bars of the Fraser? Gold had also been found in cracks in the rock along the river. Whence had it come? The man farthest upstream in spring would be on the ground first for the great find that was bound to make some seeker's fortune. So all stayed who could. Fortunately, the winter of '58–'59 was mild, the autumn late, the snowfall light, and the spring very early. Fate, as usual, favoured the dauntless.

In parties of twos and tens and twenties, and even as many as five hundred, the miners began moving up the river prospecting. Those with horses had literally to cut the way with their axes over windfall, over steep banks, and round precipitous cliffs. Where rivers had to be crossed, the men built rude rafts and poled themselves over, with their pack-horses swimming behind. Those who had oxen killed the oxen and sold the beef. Others breasted the mill-race of the Fraser in canoes and dugouts. Governor Douglas estimated that before April of '59 as many as three hundred boats with five men in each had ascended the Fraser. Sometimes the amazing spectacle was seen of canoes lashed together in the fashion of pontoon bridges, with wagons full of provisions braced across the canoes. These travellers naturally did not attempt Fraser Canyon.

Before Christmas of '59 prospectors had spread into Lillooet and up the river as high as Chilcotin, Soda Creek, Alexandria, Cottonwood Canyon, Quesnel, and Fort George. It was safer to ascend such wild streams than to run with the current, though countless canoes and their occupants were never heard of after leaving Yale. Where the turbid yellow flood began to rise and 'collect'—a boatman's phrase—the men would scramble ashore, and, by means of a long tump-line tied—not to the prow, which would send her sidling—to the middle of the first thwart, would tow their craft slowly up-stream. I have passed up and down Fraser Canyon too often to count the times, and have canoed one wild rapid twice, but never without wondering how those first gold-seekers managed the ascent in that winter of '59.

There was no Cariboo Road then. There was only the narrow footpath of the trapper and the fisherman close down to the water; and when the rocks broke off in sheer precipice, an unsteady bridge of poles and willows spanned the abyss. A 'Jacob's ladder' a hundred feet above a roaring whirlpool without handhold on either side was one thing for the Indian moccasin and quite another thing for the miner's hobnailed boot. The men used to strip at these places and attempt the rock walls barefoot; or else they cached their canoe in a tree, or hid it under moss, lashed what provisions they could to a dog's back, and, with a pack strapped to their own back, proceeded along the bank on foot. The trapper carries his pack with a strap round his forehead. The miner ropes his round under his shoulders. He wants hands and neck free for climbing. Usually the prospectors would appoint a rendezvous. There, provisions would be slung in the trees above the reach of marauding beasts, and the party would disperse at daybreak, each to search in a different direction, blazing trees as he went ahead so that he could find the way back at night to the camp. Distress or a find was to be signalled by a gunshot or by heliograph of sunlight on a pocket mirror; but many a man strayed beyond rescue of signal and never returned to his waiting 'pardners.' Some were caught in snowslides, only to be dug out years later.

Many signs guided the experienced prospector. Streams clear as crystal came, he knew, from upper snows. Those swollen at midday came from near-by snowfields. Streams milky or blue or peacock green came from glaciers—ice grinding over rock.

Heavy mists often added to the dangers. I stood at the level of eight thousand feet in this region once with one of the oldest prospectors of the canyon. He had been a great hunter in his day. A cloud came through a defile of the peaks heavy as a blanket. Though we were on a well-cut bridle-trail, he bade us pause, as one side of the trail had a sheer drop of four thousand feet in places. 'Before there were any trails, how did you make your way here to hunt the mountain goat when this kind of fog caught you?' I asked.

'Threw chips of stone ahead and listened,' he answered, 'and let me tell you that only the greenest kind of tenderfoot ever takes risks on a precipice.'

And nine men out of ten were such green tenderfoots that winter of '58–'59, when five thousand prospectors overran the wild canyons and precipices of the Fraser. Two or three things the prospector always carried with him—matches, a knife, a gun, rice, flour, bacon, and a little mallet-shaped hammer to test the 'float.' What was the 'float'? A sandy chunk of gravel perhaps flaked with yellow specks the size of a pin-head. He wanted to know where that chunk rolled down from. He knocked it open with his mallet. If it had a shiny yellow pebble inside only the size of a pea, the miner would stay on that bank and begin bench diggings into the dry bank. By the spring of '59 dry bench diggings had extended back fifty miles from the river. If the chunk revealed only tiny yellow specks, perhaps mixed with white quartz, the miner would try to find where it rolled from and would ascend the gully, or mountain torrent, or precipice. Queer stories are told of how during that winter almost bankrupt grocers grubstaked prospectors with bacon and flour and received a half-interest in a mine that yielded five or six hundred dollars a day in nuggets.

But for one who found a mine a thousand found nothing. The sensations of the lucky one beggared description. 'Was it luck or was it perseverance?' I asked the man who found one of the richest silver-mines in the Big Bend of the Columbia. 'Both and mostly dogged,' he answered. 'Take our party as a type of prospectors from '59 to '89, the thirty years when the most of the mining country was exploited. We had come up, eleven green kids and one old man, from Washington. We had roughed it in East and West Kootenay and were working south to leave the country dead broke. We had found "float" in plenty, and had followed it up ridges and over divides across three ranges of mountains. Our horses were plumb played out. We had camped on a ridge to let them fatten up enough to beat it out of British Columbia for ever. Well, we found some galena "floats" in a dry gully on the other side of the valley. We had provisions left for only eleven days. Some of the boys said they would go out and shoot enough deer to last us for meat till we could get out of the country. Old Sandy and I thought we would try our luck for just one day. We followed that "float" clear across the valley. We found more up the bed of a raging mountain torrent; but the trouble was that the stream came over a rock sheer as the wall of a house. I was afraid we'd lose the direction if we left the stream bed, but I could see high up the precipice where it widened out in a bench. You couldn't reach it from below, but you could from above, so we blazed the trees below to keep our direction and started up round the hog's back to drop to the bank under. By now it was nightfall, and we hadn't had anything to eat since six that morning. Old Sandy wanted to go back, but I wouldn't let him. He was trembling like an aspen leaf. It is so often just the one pace more that wins or loses the race. We laboured up that slope and reached the bench just at dark. We were so tired we had hauled ourselves up by trees, brushwood branches, anything. I looked over the edge of the rock. It dropped to that shelf we had seen from the gully below. It was too dark to do anything more; we knew the fellows back at the camp on the ridge would be alarmed, but we were too far to signal.'

'How far?' I asked.

'About twenty-two miles. We threw ourselves down to sleep. It was terribly cold. We were high up and the fall frosts were icy, I tell you! I woke aching at daybreak. Old Sandy was still sleeping. I thought I would let myself down over the ledge and see what was below, for there were no mineral signs where we were. I crawled over the ledge, and by sticking my fingers and toes in the rocks got down to about fifteen feet from the drop to a soft grassy level. I looked, hung for a moment, let go, and "lit" on all fours. Then I looked up! The sun had just come over that east ridge and hit the rocks. I can't talk about it yet! I went mad! I laughed! I cried! I howled! There wasn't an ache left in my bones. I forgot that my knees knocked from weakness and that we had not had a bite for twenty-four hours. I yelled at Old Sandy to wake the dead. He came crawling over the ledge and peeked down. "What's the matter?" says he. "Matter," I yelled. "Wake up, you old son of a gun; we are millionaires!" There, sticking right out of the rock, was the ledge where "float" had been breaking and washing for hundreds of years; so you see, only eleven days from the time we were going to give up, we made our find. That mine paid from the first load of ore sent out by pack-horses.'

Other mines were found in a less spectacular way. The 'float' lost itself in a rounded knoll in the lap of a dozen peaks; and the miners had to decide which of the benches to tunnel. They might have to bring the stream from miles distant to sluice out the gravel; and the largest nuggets might not be found till hundreds of feet had been washed out; but always the 'float,' the pebbles, the specks that shone in the sun, lured them with promise. Even for those who found no mine the search was not without reward. There was the care-free outdoor life. There was the lure of hope edging every sunrise. There was the fresh-washed ozone fragrant with the resinous exudations of the great trees of the forest. There was the healing regeneration to body and soul. Amid the dance-halls and saloons the miner with money becomes a sot. Out in the wilds he becomes a child of nature, simple and clean and elemental as the trees around him or the stars above him.

I think of one prospector whose range was at the headwaters of the Athabaska. In the dance-halls he had married a cheap variety actress. When the money of his first find had been dissipated she refused to live with him, and tried to extort high alimony by claiming their two-year-old son. The penniless prospector knew that he was no equal for law courts and sheriffs and lawyers; so he made him a raft, got a local trader to outfit him, and plunged with his baby boy into the wilderness, where no sheriff could track him. I asked him why he did not use pack-horses. He said dogs could have tracked them, but 'the water didn't leave no smell.' In the heart of the wilderness west of Mounts Brown and Hooker he built him a log cabin with a fireplace. In that cabin he daily hobbled his little son, so that the child could not fall in the fire. He set his traps round the mountains and hunted till the snow cleared. By the time he could go prospecting in spring he had seven hundred dollars' worth of furs to sell; and he kept the child with him in the wilds till his wife danced herself across the boundary. Then he brought the boy down and sent him to school. When the Canadian Pacific Railway crossed the Rockies, that man became one of the famous guides. He was the first guide I ever employed in the mountains.

Up-stream, then, headed the prospectors on the Fraser in that autumn of '58. The miner's train of pack-horses is a study in nature. There is always the wise old bell-mare leading the way. There is always the lazy packer that has to be nipped by the horse behind him. There are always the shanky colts who bolt to stampede where the trail widens; but even shanky-legged colts learn to keep in line in the wilds. At every steep ascent the pack-train halts, girths are tightened, and sly old horses blow out their sides to deceive the driver. At first colts try to rub packs off on every passing tree, but a few tumbles heels over head down a bank cure them of that trick.

Always the course in new territory is according to the slope of the ground. River-bank is followed where possible; but where windfall or precipice drives back from the bed of the river over the mountain spurs, the pathfinder takes his bearings from countless signs. Moss is on the north side of tree-trunks. A steep slope compels a zigzag, corkscrew ascent, but the slope of the ground guides the climber as to the way to go; for slope means valley; and in valleys are streams; and in the stream is the 'float,' which is to the prospector the one shining signal to be followed. Timber-line is passed till the forests below look like dank banks of moss. Cloud-line is passed till the clouds lie underneath in grey lakes and pools. A 'fool hen' or mountain grouse comes out and bobbles her head at the passing pack-train. A whistling marmot pops up from the rocks and pierces the stillness. Redwings and waxbills pick crumbs from every camp meal; and occasionally a bald-headed eagle utters a lonely raucous cry from solitary perch of dead branch or high rock.

Naturally enough, the pack-train unconsciously follows the game-trail of deer and goat and cougar and bear across the slope to the watering-places where springs gush out from the rocks. One has only to look close enough to see the little cleft footprint of the deer round these springs. To the miners, penetrating the wilds north of the Fraser, the caribou proved a godsend during that lean first winter. The miners spelled it 'cariboo,' and thus gave the great gold area its name.

The population of Yale that winter consisted of some eight hundred people, housed in tents and log shacks roofed with canvas. Between Yale and Hope remained two thousand miners during the winter. Meals cost a dollar, served on tin plates to diners standing in long rows waiting turn at the counter. The regular menu at all meals was bacon, salmon, bread, and coffee. Of butter there was little; of milk, none. Wherever a sand-bar gave signs of mineral, it was tested with the primitive frying-pan. If the pan showed a deposit, the miner rigged up a rocker—a contraption resembling a cradle with rockers below, about four feet from end to end, two feet across, and two deep. The sides converged to bottom. At the head was a perforated sheet-iron bottom like a housewife's colander. Into this box the gravel was shovelled by one miner. The man's 'pardner' poured in water and rocked the cradle—cradled the sand. The water ran through the perforated bottom to a second floor of quicksilver or copperplate or woolly blanket which caught the gold. On a larger scale, when streams were directed through wooden boxes, the gold was sluiced; on a still larger scale, the process was hydraulic mining, though the same in principle. In fact, in huge free milling works, where hydraulic machinery crushes the gold-bearing quartz and screens it to fineness before catching the gold on delicate sieves, the process is only a com-plex refinement of the bar-washer cradling his gold.

In the Rocky Mountains
FROM A PHOTOGRAPH

Fires had not yet cleared the giant hemlock forests, as they have to-day along the Cariboo Trail, and prospectors found their way through a chartless sea of windfall—hemlocks criss-crossed the height of a house with branches interlaced like wire. Cataracts fell over lofty ledges in wind-blown spray. Spanish moss, grey-green and feathery, hung from branch to branch of the huge Douglas firs. Sometimes the trail would lead for miles round the edge of some precipices beyond which could be glimpsed the eternal snows. Sometimes an avalanche slid over a slope with the distant appearance of a great white waterfall and the echo of muffled thunder. Where the mountain was swept as by a mighty besom, the pack-train kept an anxious eye on the snow amid the valleys of the upper peaks; for, in an instant, the snowslide might come over the edge of the upper valley to sweep down the slope, carrying away forests, rocks, trail, pack-train and all. The story is told of one slide seen by the guide at the head of a long pack-train. He had judged it to be ten miles away; but out from the upper valley it came coiling like a long white snake, and before he could turn, it had caught him. In a slide death was almost certain, from suffocation if not from the crush of falling trees and rocks. Miners have been taken from their cabins dead in the trail of a snowslide that swept the shack to the bottom of the valley without so much as a hair of their heads being injured. Though the logs were twisted and warped, the dead bodies were not even bruised.

When a hushed whisper came through the trees, travellers looked for some waterfall. At midday, when the thaw was at its full, all the mountain torrents became vocal with the glee of disimprisoned life running a race of gladness to the sea. The sun sets early in the mountains with a gradual hushing of the voice of glad waters and a red glow as of wine on the encircling peaks. Camp for the night was always near water for the horses; and every star was etched in replica in river or lake. Sunrise steals in silence among the mountain peaks. There is none of that stir of song and vague rustling of animal life such as are heard at lower levels. Nor does the light gradually rise above the eastern horizon. The walled peaks cut off the skyline in mid-heaven. The stars pale. Trees and crags are mirrored in the lake so clearly that one can barely tell which is real and which is reflection. Then the water-lines shorten and the rocks emerge from the belts and wisps of mist; and all the sunset colours of the night before repeat themselves across the changing scene. As you look, the clouds lift. The cook shouts 'breakfast!' And it is another day.

Such was the trail and the life of the prospector who beat his way by pack-train and canoe up the canyons of the Fraser to learn whence came the wash of gold flake and nugget which he found in the sand-bars below.

CHAPTER THREE

CARIBOO

Indian unrest was probably first among the causes which led the miners to organize themselves into leagues for protection. The Indians of the Fraser were no more friendly to newcomers now than they had been in the days of Alexander Mackenzie and Simon Fraser.[1] They now professed great alarm for their fishing-grounds. Men on the gold-bars were jostled and hustled, and pegs marking limits were pulled up. A danger lay in the rows of saloons along the water-front—the well-known danger of liquor to the Indian. So the miners at Yale formed a vigilance committee and established self-made laws. The saloons should be abolished, they decreed. Sale of liquor to any person whomsoever was forbidden. All liquor, wherever found, was ordered spilled. Any one selling liquor to an Indian should be seized and whipped thirty-nine lashes on the bare back. A standing committee of twelve was appointed to enforce the law till the regular government should be organized.

It was July '58 when the miners on the river-bars formed their committee. And they formed it none too soon, for the Indians were on the war-path in Washington and the unrest had spread to New Caledonia. Young McLoughlin, son of the famous John McLoughlin of Oregon, coming up the Columbia overland from Okanagan to

Kamloops with a hundred and sixty men, four hundred pack-horses and a drove of oxen, had three men sniped off by Indians in ambush and many cattle stolen. At Big Canyon on the Fraser two Frenchmen were found murdered. When word came of this murder the vigilance committee of Yale formed a rifle company of forty, which in August started up to the forks at Lytton. At Spuzzum there was a fight. Indians barred the way; but they were routed and seven of them killed in a running fire, and Indian villages along the river were burned. Meanwhile a hundred and sixty volunteers at Yale formed a company to go up the river under Captain Snyder. The company's trader at Yale was reluctant to supply arms, for the company's policy had ever been to conciliate the Indians. But, when a rabble of two thousand angry miners gathered round the store, the rifles were handed over on condition that forty of the worst fire-eaters in the band should remain behind. Snyder then led his men up the river and joined the first company at Spuzzum. At China Bar five miners were found hiding in a hole in the bank. With a number of companions they had been driven down-stream from the Thompson by Indians and had been sniped all the way for forty miles. Man after man had fallen, and the five survivors in the bank were all wounded.

When the Indians saw the company of armed men under Snyder, they fled to the hills. Flags of truce were displayed on both sides and a peace was patched up till Governor Douglas could come up from the coast. Not, however, before there occurred an unfortunate incident. At Long Bar, when an Indian chief came with a flag of truce, two of the white men snatched it from him and trampled it in the mud. On the instant the Indians shot both the white men where they stood.

Douglas had been up as far as Yale in June, but was now back in Victoria, where couriers brought him word of the open fight in August. He promptly organized a force of Royal Engineers and marines and set out for the scene of the disorders. Royal Engineers to the number of a hundred and fifty-six and their families had come out from England for the boundary survey; and their presence must have seemed providential to Douglas, now that the miners were forming vigilance committees of their own and the Indians were on the war-path. He went up the river in a small cruiser and reached Hope on the 1st of September. Salutes were fired as he landed. Douglas knew how to use all the pomp of regimentals and formality to impress the Indians. He opened a solemn powwow with the chiefs of the Fraser. As usual, the white man's fire-water was found to be the chief cause of the trouble. Without waiting for legislative authority, Douglas issued a royal proclamation against the sale of liquor and left a mining recorder to register claims. He also appointed a justice of the peace. Then he went on to Yale. At Yale he considered the price of provisions too high, and by arbitrarily reducing the price at the company's stores, he broke the ring of the petty dealers. This won him the friendship of the miners. Within a week he had allayed all irritation between white man and Indian. In a quarrel over a claim a white man had been murdered on one of the bars. Douglas appointed magistrates to try the case. The trial was of course illegal, for colonial government had not been formally inaugurated in New Caledonia or British Columbia, as it was soon to be known, and Douglas's authority as governor did not extend beyond Vancouver Island. But so, for that matter, were illegal all his actions on this journey; yet by an odd inconsistency of fact against law, they restored peace and order on the river.

It was not long, however, before the formal organization of the new colony took place. Hardly had Douglas returned to Victoria when ships from England arrived bringing his commission as governor of British Columbia. Arrived, also, Matthew Baillie Begbie, 'a Judge in our Colony of British Columbia,' and a detachment of Royal Engineers under command of Colonel Moody. At Fort Langley, on November 19, 1858, the colony of British Columbia was proclaimed under the laws of England.

Then, in January, just as Douglas and the officers of his government had again settled down comfortably at Victoria, came word of more riots at Yale, led by a notorious desperado and deposed judge of California

A group of Thompson River Indians
FROM A PHOTOGRAPH BY MAYNARD

named Ned McGowan. The possibility of American occupation
had become an obsession at Victoria. There were undoubtedly those
among the American miners who made wild boasts. Douglas gathered
up all his panoply of war and law. Along went Colonel Moody, with
a company of his Royal Engineers, Lieutenant Mayne of the Imperial
Navy with a hundred bluejackets, and Judge Matthew Begbie, to
deal out justice to the offenders. Douglas remembered the cry 'fifty-
four forty or fight,' and he remembered what had happened to his
chief, McLoughlin, in Oregon when the American settlers there had
set up vigilance committees. He would take no chances. The party
carried along a small cannon. Lieutenant Mayne could not take his
cruiser the *Plumper* higher than Langley; and there the forces were
transferred to Tom Wright's stern-wheeler, the *Enterprise*. But, when
they arrived at Hope, the whole affair looked like semi-comic vaude-
ville. Yale, too, was as quiet as a church prayer-meeting; and Colonel
Moody preached a sermon on Sunday to a congregation of forty in
the court-house—the first church service ever held on the mainland
of British Columbia.

The trouble had happened in this way. Christmas Day had been celebrated hilariously. At Yale a miner of Hill's Bar, some miles down the river, had beaten up a negro. The Yale magistrate had issued a warrant for the miner's arrest—poor magistrate, he had found little to do since his appointment in September! The miner, now sobered, fled back to his bar. The warrant was sent after him to the local peace officer for execution, but this officer had already issued a warrant for the arrest of the negro at Yale; so there it stood—each fighter making complaint against the other and the two magistrates in lordly contempt of each other! The man who tried to arrest the negro was insolent and was jailed by the Yale magistrate. Ned McGowan, the Californian down on the bar, then came up to Yale with a posse of twenty men to arrest the magistrate for arresting the man who had been sent to arrest the negro. Bursting with rage, the astonished dignitary at Yale was bundled into a canoe. He was fined fifty dollars for contempt of court.

It was at this stage of the comedy of errors that Moody, Begbie, and Mayne came on the scene. At first McGowan showed truculence and assailed Moody; but when he saw the force of engineers and bluejackets and saw the big gun hoisted ashore, he apologized, paid his fine for the assault, and invited the officers to a champagne dinner on Hill's Bar. Both sides to the quarrel cooled down and the riots ended. The army stayed only to see the miners wash the gold and then put back to Victoria. The miners had learned that an English judge and a field force could be put on the ground in a week. September had settled disorder among the Indians. January settled disorder among the whites.

In the wild remote regions of the up-country there was much 'claim jumping.' A man lost his claim if he stopped mining for seventy-two hours, and when the place of registration was far from the find, 'pardners' camped on the spot in dugouts or in lean-tos of logs and moss along the river-bank. There were fights and there was killing, and sometimes the river cast up its dead. The marvel is that there were not more crimes. In every camp is a species of human

Sir Matthew Baillie Begbie
FROM A PORTRAIT BY SAVANNAH

vulture living off other men's risk. Whenever a lone man came in from the hills and paid for his purchase in nuggets, such vultures would trail him back to his claim and make what they could out of his discovery.

So, by pack-train and canoe, the miners worked up to Alexandria, to Quesnel, to Fort George. Towards spring, when the prospectors had succeeded in packing in more provisions, they began striking back east from the main river, following creeks to their sources, and from their sources over the watershed to the sources of creeks flowing in an opposite direction. Late in '59 men reached Quesnel Lake and Cariboo Lake. Binding saplings together with withes, the prospectors poled laboriously round these alpine lagoons, and where they found creeks pouring down from the upper peaks, they followed these creeks up to their sources. Pockets of gravel in the banks of both lakes yielded as much as two hundred dollars a day. On Horse Fly Creek up from Quesnel Lake five men washed out in primitive rockers a hundred ounces of nuggets in a week. The gold-fever, which had subsided when all the bars of the Fraser were occupied, mounted again. Great rumours began to float out from the up-country. Bank facings seemed to indicate that the richest pay-dirt lay at bed-rock. This kind of mining required sluicing, and long ditches were constructed to bring the water to the dry diggings. By the autumn of '59 a thousand miners were at work round Quesnel Lake. By the spring of '60 Yale and Hope were almost deserted. Men on the upper diggings were making from sixty to a hundred dollars a day. Only Chinamen remained on the lower bars.

It was in the autumn of the year '60 that Doc Keithley, John Rose, Sandy MacDonald, and George Weaver set out from Keithley Creek, which flows into Cariboo Lake, to explore the cup-like valley amid the great peaks which seemed to feed this lake. They toiled up the creek five miles, then followed signs up a dry ravine seven miles farther. Reaching the divide at last, they came on an open park-like ridge, bounded north and east by lofty shining peaks. Deer and caribou tracks were everywhere. It was now that the region became known as Cariboo. They camped on the ridge, cooked supper, and slept under the stars. Should they go on, or back? This was far above the benches of wash-gravel. Going up one of the nameless peaks, they stepped out on a ledge and viewed the white, silent mountain-world. Marmots stabbed the lonely solitude with echoing whistle. Wind came up from the valley in the sibilant sigh of a sea. It was doubtful if even Indians had ever hunted this ground. The game was so tame, it did not know enough to be afraid. The men could see another creek shining in the sunrise on the other side of the ridge. It seemed to go down to a valley benched by gravel flanks. They began wandering down that creek and testing the gravel. Before they had gone far their eyes shone like the wet pebbles in their hands. The gravel was pitted with little yellow stones. Where rain and spring-wash had swept off the gravel to naked rock, little nuggets lay exposed. The men began washing the gravel. The first pan gave an ounce; the second pan gave nuggets to the weight of a quarter of a pound. The excited prospectors forgot time. Dark was falling. They slept under their blankets and awoke at daybreak below twelve inches of snow.

They were out of provisions. Somebody had to go back down to Cariboo Lake for food. Each man staked out a claim. And, while two built a log cabin, the other two set off over the hills for food. There was some sort of a log store down at Cariboo Lake. The one thing these prospectors were determined on was secrecy till they could get their claims registered. Bands of nondescript men hung round the provision-store of Cariboo Lake awaiting a breath to fan their flaming hopes of fortune. What let the secret out at the store is not known. Perhaps too great an air of secrecy. Perhaps too strenuous denials. Perhaps the payment of provisions in nuggets. But when these two packed back over the hills on snowshoes, they were trailed. Followers came in with a whoop behind them on Antler Creek. Claims were staked faster than they could be recorded. The same claims were staked over and over, the corner of one overlapping another. When the gold commissioner came hurriedly across the country in March, he found the MacDonald-Rose party living in a cabin and the rest of the camp holding down their claims by living in holes which they had dug in the ground.

This was the spring of '61; and Antler Creek proved only the beginning of the rush to Cariboo. Over the divide in mad stampede rushed the gold-seekers northward and eastward. Ed Stout and Billy Deitz and two others found signs that seemed very poor on a creek which they named William's after Deitz. The gold did not pan a dollar a wash; but in wild haste came the rush to William's Creek. Crossing a creek one party of prospectors was overtaken by a terrific thunderstorm, with rock-shattering flashes of lightning. Shivering in the canyon, but afraid to stand under trees or near rocks, with the gravel shelving down all round them, one of the men exclaimed sardonically, 'Well, boys, this *is* lightning.' The stream became known as Lightning Creek and proved one of the richest in Cariboo. William's Creek was panning poorer and poorer and was being called 'Humbug Creek,' when miners staked near by decided to see what they could find beneath the blue clay. It took forty-eight hours to dig down. The reward was a thousand dollars' worth of wash-gravel. Back surged the miners to William's Creek. They put shafts and tunnels through the clay and sluiced in more water for hydraulic work. Claims on William's Creek produced as high as forty pounds of gold in a day. From another creek, only four hundred feet long, fifty thousand dollars' worth of gold was washed within a space of six weeks. Lightning Creek yielded a hundred thousand dollars in three weeks. In one year gold to the value of two and a half million dollars was shipped from Cariboo.

Millions were not so plentiful in those days, and the reports which reached the outside world sounded like the *Arabian Nights* or some fairy-tale. The whole world took fire. Cariboo was on every man's lips, as were Transvaal and Klondike half a century later. The New England States, Canada, the Maritime Provinces, the British Isles—all were set agog by the reports of the new gold-camps where it was only necessary to dig to find nuggets. By way of Panama, by way of San Francisco, by way of Spokane, by way of Victoria, by way of Winnipeg and Edmonton came the gold-seekers, indifferent alike to perils of sea and perils of mountain. Men who had never seen a mountain thought airily that they could climb a watershed in a day's walk. Men who did not know a canoe from a row-boat essayed to run the maddest rapids in America. People without provisions started blindly from Winnipeg across the width of half a continent. In the mad rush were clerks who had never seen 'float,' English school-teachers whose only knowledge of gold was that it was yellow, and dance-hall girls with very little possession of anything on earth but recklessness and slippers; and the recklessness and the slippers danced them into Cariboo, while many a solemn wight went to his death in rockslide or rapids. By the opening of '62 six thousand miners were in Cariboo, and Barkerville had become the central camp. How these people ever gained access to the centre of the wilderness before the famous Cariboo Road had been built is a mystery. Some arrived by pack-train, some by canoe, but the majority afoot.

Governor Douglas could not regulate prices here, and they jumped to war level. Flour was three hundred dollars a barrel. Dried apples brought two dollars and fifty cents a pound; and for lack of fruit many miners died from scurvy. Where gold-seekers tramped six hundred miles over a rocky trail, it is not surprising that boots commanded fifty dollars a pair. Of the disappointed, countless numbers filled unknown graves, and thousands tramped their way out starving and begging a meal from the procession of incomers.

The places of the gold deposits were freakish and unaccountable. Sometimes the best diggings were a mother lode at the head of a creek. Sometimes they were found fifty feet under clay at the foot of a creek where the dashing waters swerved round some rocky point into a river. Old miners now retired at Yale and Hope say that the most ignorant prospector could guess the place of the gold as well as the geologist. Billy Barker, after whom Barkerville was named, struck it rich by going fifty feet below the surface down the canyon. Cariboo Cameron, the luckiest of all the miners and not originally a prospector, found his wealth by going still lower on the watercourse to a vertical depth of eighty feet.

For seven miles along William's Creek worked four thousand men. Cariboo Cameron took a hundred and fifty thousand out of his claim in three months. In six months of '63 William's Creek yielded a million and a half dollars, and this was only one of many rich creeks. From '59 to '71 came twenty-five million dollars in gold from the Cariboo country. By '65 hydraulic machinery was coming in and the prospectors were flocking out; but to this day the Cariboo mines have remained a freakish gamble. Mines for which capitalists have paid hundreds of thousands have suddenly ended in barren rock. Diggings from which nuggets worth five hundred dollars have been taken have petered out after a few hundred feet. Even where the gravel merged to whitish gold quartz, the most expert engineer in the camp could not tell when the vein would fault and cease as entirely as if cut off. And the explanation of this is entirely theoretical. The theory is that the place of the gold was the gravel bed of an old stream, an old stream antedating the petrified forests of the South-west, and that, when vast alluvial deposits were carried over a great part of the continent by inland lakes and seas, the gold settled to the bottom and was buried beneath the deposits of countless centuries. Then convulsive changes shook the earth's surface. Mountains heaved up where had been sea bottom and swamp and watery plain. In the upheaval these subterranean creek beds were hoisted and thrown towards the surface. Floods from the eternal snows then grooved out watercourses down the scarred mountainsides. Frost and rain split away loose debris. And man found gold in these prehistoric, perhaps preglacial, creek beds. However this may be, there was no possible scientific way of knowing how the gold-bearing area would run. A fortune might come out of one claim of a hundred feet and its next-door neighbour might not yield an atom of gold. Only the genii of the hidden earth held the secret; and modern science derides the invisible pixies of superstition, just as these invisible spirits of the earth seem to laugh at man's best efforts to ferret out their secrets.

What became of the lucky prospectors? I have talked with some of them on the lower reaches of the Cariboo Road. They are old and poor to-day, and the memory of their fortune is as a dream. Have they not lived at Hope and Yale and Lytton for fifty years and seen their trail crumble into the canyon, with not a dozen pack-trains a year passing to the Upper country? John Rose, who was one of the men to find Cariboo, set out in the spring of '63 to prospect the Bear River country. He set out alone and was never again seen alive. Cariboo Cameron, a 'man from Glengarry,' went back to Glengarry by the Ottawa and established something like a baronial estate; but he lost his money in various investments and died in 1888 in Cariboo a poor man. Billy Deitz, after whom a famous creek was named, died penniless in Victoria; and the Scottish miner who rhymed the songs of Cariboo died unwept and unknown to history.

The romance of the trail is almost incredible to us, who may travel by motor from Ashcroft to Barkerville. In October '62 a Mr Ireland and a party were on the trail when snow began falling so heavily that it was unsafe to proceed. They halted at a negro's cabin. Out of the heavy snowfall came another party struggling like themselves. Then a packer emerged from the storm with word that five women and twenty-six men were snowbound half a mile ahead. Ireland and his party set out to the rescue; but they lost the trail and could only find the cabin again by means of the gunshots that the others kept firing as a signal. Two dozen people slept that night in the log shack; and when dawn came, four feet of snow lay on the ground and the great evergreens looked like huge sugar-cones. On snowshoes Ireland and three others set out to find the lost men and women on the lower trail. They found them at sundown camped in a ravine beside a rock, with their blankets up to keep off the wind, thawing themselves out before a fire. A high wind was blowing and it was bitterly cold. The lost people had not eaten for three days. Twenty men from the cabin dug a way through the drifts with their snowshoes and brought horses to carry the women back to the coloured man's roof.

But it was not of the perils of the trail that the outside world heard. The outside world heard of claims which any man might find and from which gold to the value of a hundred and fifty thousand dollars could be dug and washed in three months. The outside world thought that gold could be picked up amid the rocks of British Columbia. Necessity is the mother of invention. She is also the hard foster-mother of desperation and folly. Times were very hard in Canada. The East was hard up. Farming did not pay. All eyes turned towards Cariboo; and no wonder! Many of the treasure-seekers holding the richest claims had gone to Cariboo owning nothing but the clothes on their backs. A season's adventure in a no-man's-land of bear and deer, above cloud-line and amid wild mountain torrents, had sent them out to the world laden with wealth. Some ran the wild canyons of the Fraser in frail canoes and crazy rafts with their gold strapped to their backs or packed in buckskin sacks and carpet-bags. And some who had won fortune and were bringing it home went to their graves in Fraser Canyon.

CHAPTER FOUR

THE OVERLANDERS

When the Cariboo fever reached the East, the public there had heard neither of the Indian massacres in Oregon nor that the Sioux were on the war-path in Dakota. Promoters who had never set foot west of Buffalo launched wild-cat mining companies and parcel express devices and stages by routes that went up sheer walls and crossed unbridged rivers. To such frauds there could be no certain check; for it took six months to get word in and out of Cariboo. Eastern papers were full of advertisements of easy routes to the gold-diggings. Far-off fields look green. Far-off gold glittered the brighter for the distance. Cariboo became in popular imagination a land where nuggets grew on the side of the road and could be picked by the bushel-basket. Besides, times were so hard in the East that the majority of the youthful adventurers who were caught by the fever had nothing to lose except their lives.

A group of threescore young men from different parts of Canada, from Kingston, Niagara, and Montreal, having noticed advertisements of an easy stage-route from St Paul, set out for the gold-diggings in May 1862. Tickets could be purchased in London, England, as well as in Canada, for when these young Canadians reached St Paul, they found eighteen young men from England, like themselves, diligently searching the whereabouts of the stage-route. That was their first inkling that fraudulent practices were being carried on and that they had been deceived, that there was, in fact, no stage-route from St Paul to Cariboo. A few of them turned back, but the majority, by ox-cart and rickety stagecoach, pushed on to the Red River and went up to a point near the boundary of modern Manitoba, where lay the first steamboat to navigate that river, about to start on her maiden trip. On this steamboat, the little *International*, afterwards famous for running into sand-banks and mud-bars, the troops of Overlanders took passage, and stowed themselves away wherever they could, some in the cook's galley and some among the cordwood piled in the engine-room.

The Sioux were on a rampage in Minnesota and Dakota, but Alexander Dallas, governor of Rupert's Land for the Hudson's Bay Company, and Mgr Taché, bishop of St Boniface, were aboard, and their presence afforded protection. On the way to the vessel some of the Overlanders had narrowly escaped a massacre. The story is told that as they slowly made their way in ox-carts up the river-bank, a band of horsemen swept over the horizon, and the travellers found themselves surrounded by Sioux warriors. The old plainsman who acted as guide bethought him of a ruse: he hoisted a flag of the Hudson's Bay Company and waved it in the face of the Sioux without speaking. The painted warriors drew together and conferred. The oxen stood complacently chewing the cud. Indians never molested British fur-traders. Presently the raiders went off over the horizon as swiftly as they had come, and the gold-seekers drove on, little realizing the fate from which they had been delivered.

There had been heavy rains that spring on the prairie, and trees came jouncing down the muddy flood of the Red River. The little *International*, like a panicky bicycle rider, steered straight for every tree, and hit one with such impact that her smokestack came toppling down. At another place she pushed her nose so deep in the soft mud of the riverbank that it required all the crew and most of the passengers to shove her off. But everybody was jubilant. This was the first navigation of the Red River by steam. The Queen's Birthday,

A Red River cart
FROM A PHOTOGRAPH

the 24th of May, was celebrated on board the vessel pottle-deep to
the tune of the bagpipes played by the governor's Scottish piper. But
the governor's wife was heard to lament to Bishop Taché that the
International's menu consisted only of pork and beans alternated
with beans and pork, that the service was on tin plates, and that the
dining-room chairs were backless benches.

The arrival of the steamer at Fort Garry (Winnipeg) was celebrated
with great rejoicing. Indians ran along the river-bank firing off rifles in
welcome, and opposite the flats where the fort gate opened, on what
is now Main Street, the company's men came out and fired a royal
salute. The people bound for Cariboo camped on the flats outside
Fort Garry. Here was a strange world indeed. Two-wheeled ox-carts,
made wholly of wood, without iron or bolt, wound up to the fort from
St Paul in processions a mile long, with fat squaws and whole Indian
families sitting squat inside the crib-like structure of the cart. Men and
boys loped ahead and abreast on sinewy ponies, riding bareback or on
home-made saddles. Only a few stores stood along what is now Main
Street, which ran northward towards the Selkirk Settlement. With
the Indians, who were camped everywhere in the woods along the

Assiniboine, the Overlanders began to barter for carts, oxen, ponies, and dried deer-meat or pemmican. An ox and cart cost from forty to fifty dollars. Ponies sold at twenty-five dollars. Pemmican cost sixteen cents a pound, and a pair of duffel Hudson's Bay blankets cost eight or ten dollars. Instead of blankets, many of the travellers bought the cheaper buffalo robes. These sold as low as a dollar each.

John Black, the Presbyterian 'apostle of the Red River,' preached special sermons on Sunday for the miners. And on a beautiful June afternoon the Overlanders headed towards the setting sun in a procession of almost a hundred ox-carts; and the fort waved them farewell. One wonders whether, as the last ox-cart creaked into the distance, the fur-traders realized that the miner heralded the settler, and that the settler would fence off the hunter's game preserve into farms and cities. A rare glamour lay over the plains that June, not the less rare because hope beckoned the travellers. The unfenced prairie billowed to the horizon a sea of green, diversified by the sky-blue waters of slough and lake, and decked with the hues of gorgeous flowers—the prairie rose, fragrant, tender, elusive, and fragile as the English primrose; the blood-red tiger-lily; the brown windflower with its corn-tassel; the heavy wax cups of the sedgy water-lily, growing where wild duck flackered unafraid. Game was superabundant. Prairie chickens nestled along the single-file trail. Deer bounded from the poplar thickets and shy coyotes barked all night in the offing. Night in June on the northern prairie is but the shadowy twilight between two long days. The sun sets between nine and ten, and rises between three and four, and the moonlight is clear enough on cloudless nights for campers to see the time on their watches.

The trail followed was the old path of the fur-trader from fort to fort 'the plains across' to the Rockies. From the Assiniboine the road ran northerly to Forts Ellice and Carlton and Pitt and Edmonton.[1] Thomas McMicking of Niagara acted as captain and eight others as lieutenants. A scout preceded the marchers, and at sundown camp was formed in a big triangle with the carts as a stockade, the animals tethered or hobbled inside. Tents were pitched outside with six men doing sentry duty all night. At two in the morning a halloo roused camp. An hour was permitted for harnessing and breaking camp, and then the carts creaked out in line. They halted at six for breakfast and marched again at seven. Dinner was at two, supper at six, and tents were seldom pitched before nine at night. On Sunday the procession rested and some one read divine service. The oxen and ponies foraged for themselves. By limiting camp to five hours, in spite of the slow pace of the oxen, forty to fifty miles a day could be made on a good trail in fair weather. While the scout led the way, the captain and his lieutenants kept the long procession in line; and the travellers for the most part dozed lazily in their carts, dreaming of the fortunes awaiting them in Cariboo. Some nights, when the captain permitted a longer halt than usual and when camp-fires blazed before the tents, men played the violin and sang and danced. Each man was his own cook. Three or four occupied each tent. In the company was one woman, with two children. She was an Irishwoman; but she bore the name of Shubert, from which we may infer that her husband was not an Irishman.

Sunday having intervened, the travellers did not reach Portage la Prairie until the fourth day out. Another week passed before they arrived at Fort Ellice. Heavy rains came on now, and James McKay, chief trader at Fort Ellice, opened his doors to the gold-seekers. Harness and carts repaired and more pemmican bought, the travellers crossed the Qu'Appelle river in a Hudson's Bay scow, paying toll of fifty cents a cart. From the Qu'Appelle westward the journey grew more arduous. The weather became oppressively hot and mosquitoes swarmed from the sloughs. At Carlton and at Fort Pitt the fur-traders' 'string band'—husky-dogs in wolfish packs—surrounded the camp of the Overlanders and stole pemmican from under the tent-flaps. From Fort Pitt westward the trail crossed a rough, wooded country, and there were no more scows to take the ox-carts across the rivers. Eleven days of continuous rain had flooded the sloughs into swamps; and in three days as many as eight corduroy bridges had to be built. Two long trees were felled parallel and light poles were laid across

Washing gold on the Saskatchewan
FROM A PHOTOGRAPH

the floating trees. Where the trees swerved to the current, some one would swim out and anchor them with ropes till the hundred carts had passed safely to the other side.

It was the 21st of July when the travellers came out on the high banks of the North Saskatchewan, flowing broad and swift, opposite Fort Edmonton. There had been floods and all the company's rafts had been carried away. But the ox-carts were poled across by means of a big York boat; and the travellers were welcomed inside the fort.

The arrival of the Overlanders is remembered at Edmonton by some old-timers even to this day. Salvoes of welcome were fired from the fort cannon by a half-breed shooting his musket into the touch-hole of the big gun. Concerts were given, with bagpipes, concertinas, flutes, drums, and fiddles, in honour of the far-travellers. Pemmican-bags were replenished from the company's stores.

Miners often uttered loud complaints against the charges made by the fur-traders for provisions, forgetting what it cost to pack these provisions in by dog-train and canoe. If the Hudson's Bay officials at Fort Garry and Edmonton had withheld their help, the Overlanders would have perished before they reached the Rockies. Though the miner did everything to destroy the fur trade—started fires which ravaged the hunter's forest haunts, put up saloons which demoralized the Indians, built wagon-roads where aforetime wandered only the shy creatures of the wilds—though the miner heralded the doom of the fur trade—yet with an unvarying courtesy, from Fort Garry to the Rockies, the Hudson's Bay men helped the Overlanders.

The majority of the travellers now changed oxen and carts for pack-horses and *travois*, contrivances consisting of two poles, within which the horses were attached, and a rude sledge. A few continued with oxen, and these oxen were to save their lives in the mountains.

The farther the Overlanders now plunged into the wilderness, the more they were pestered by the husky-dogs that roamed in howling hordes round the outskirts of the forts. The story is told of several prospectors of this time, who slept soundly in their tent after a day's exhausting tramp, and awoke to find that their boots, bacon, rope, and clothes had been devoured by the ravenous dogs. They asked the trader's permission to sleep inside the fort.

'Why?' asked the amused trader. 'Why, now, when the huskies have chewed all you own but your instruments? You are locking the stable door after your horse has been stolen.'

'No,' answered the prospectors. 'If those husky-dogs last night could devour all our camp kit without disturbing us, to-night they might swallow us before we'd waken.'

The next pause was at St Albert, one of Father Lacombe's missions. What surprised the Overlanders as they advanced was the amazing fertility of the soil. At Fort Garry, at Pitt, at Edmonton, at St Albert, at St Ann, they saw great fields of wheat, barley, and potatoes. Afterwards many who failed in the mines drifted back to the plains and became farmers. The same thing had happened in California, and was repeated at a later day in the rush to the Klondike. Great seams of coal, too, were seen projecting from the banks of the

In the Yellowhead Pass
FROM A PHOTOGRAPH

Saskatchewan. Here some of the men began washing for gold, and, finding yellow specks the size of pin-heads in the fine sand, a number of them knocked up cabins for themselves and remained west of Edmonton to try their luck. Later, when these belated Overlanders decided to follow on to Cariboo, they suffered terrible hardships.

The Overlanders were to enter the Rockies by the Yellowhead Pass, which had been discovered long ago by Jasper Hawse, of the Hudson's Bay Company. This section of their trail is visible to the modern traveller from the windows of a Grand Trunk Pacific Railway train, just as the lower sections of the Cariboo Trail in the Fraser Canyon are to be seen from the trains of the Canadian Pacific and the Canadian Northern. First came the fur-trader, seeking adventure through these passes, pursuing the little beaver. The miner came next, fevered to delirium, lured by the siren of an elusive yellow goddess. The settler came third, prosaic and plodding, but dauntless too. And then came the railroad, following the trail which had been beaten hard by the stumbling feet of pioneers.

Upper McLeod River
FROM A PHOTOGRAPH

At St Ann a guide was engaged to lead the long train of pack-
horses through the pass from Jasper House on the east to Yellowhead
Lake on the west. Colin Fraser, son of the famous piper for Sir
George Simpson of the Hudson's Bay Company, danced a Highland
fling at the gate of the fort to speed the departing guests. And to the
skirl of the bagpipes the procession wound away westward bound for
the mountains.

Instead of the thirty miles a day which they had made farther east,
the travellers were now glad to cover ten miles a day. Fallen trees lay
across the trail in impassable ramparts and floods filled the gullies.
Scouts went ahead blazing trees to show the way. Bushwhackers followed, cutting away windfall and throwing logs into sloughs. Horses
sank to their withers in seemingly bottomless muskegs,[2] so that packs
had to be cut off and the unlucky bronchos pulled out by all hands
straining on a rope.

Somewhere between the rivers Pembina and McLeod the travellers were amazed to see what the wise ones in the party thought a

volcano—a continuous and self-fed fire burning on the crown of a hill. Science of a later day pronounced this a gas well burning above some subterranean coal seam.

At length the Overlanders were ascending the banks of the McLeod, whose torrential current warned them of rising ground. Three times in one day windfall and swamp forced the party to ford the stream for passage on the opposite side. The oxen swam and the ox-carts floated and the packs came up the bank dripping. For eleven days in August every soul of the company, including Mrs Shubert's babies, travelled wet to the skin. At night great log fires were kindled and the Overlanders sat round trying to dry themselves out. Then the trail lifted to the foothills. And on the evening of the 15th of August there pierced through the clouds the snowy, shining, serrated peaks of the Rockies.

A cheer broke from the ragged band. Just beyond the shining mountains lay—Fortune. What cared these argonauts, who had tramped across the width of the continent, that the lofty mountains raised a sheer wall between them and their treasure? Cheer on cheer rang from the encampment. Men with clothes in tatters pitched caps in air, proud that they had proved themselves kings of their own fate. It is, perhaps, well that we have to climb our mountains step by step; else would many turn back. But there were no faint-hearts in the camp that night. Even the Irishwoman's two little children came out and gazed at what they could not understand.

The party now crossed a ravine to the main stream of the Athabaska. It was necessary to camp here for a week. A huge raft was built of pine saplings bound together by withes. To the stern of this was attached a tree, the branch end dipping in the water, as a sweep and rudder to keep the craft to its course. On this the Overlanders were ferried across the Athabaska. And so they entered the Yellowhead Pass.

CHAPTER FIVE

CROSSING THE MOUNTAINS

Like many lowland dwellers, the Overlanders had thought of a pass as a door opening through a rock wall. What they found was a forested slope flanked on both sides by mighty precipices down which poured cataracts with the sound of the voice of many waters. Huge hemlocks lay criss-crossed on the slope. Above could be seen the green edge of a glacier, and still higher the eternal snows of the far peaks. The tang of ice was in the air; but in the valleys was all the gorgeous bloom of midsummer—the gaudy painter's brush, the shy harebell, the tasselled windflower, and a few belated mountain roses. Long-stemmed, slender cornflowers and bluebells held up their faces to the sun, blue as the sky above them. Everywhere was an odour as of incense, the fragrance of the great hemlocks, of grasses frost-touched at night and sunburnt by day, of the unpolluted earth-mould of a thousand years.

Where was the trail? None was visible! The captain led the way, following blazes chipped in the bark of the trees, zigzagging up the slope from right to left, from left to right, hanging to the horse's mane to lift weight from the saddle, with a rest for breathing at each turn as they climbed; and, when the ridge of the foothill was surmounted, a world of peacock-blue lakes lay below, fringed by forests. The cataracts looked like wind-blown ribbons of silver.

Instead of dipping down, the trail led to the rolling flank of another great foothill, and yet another, round sharp saddlebacks connecting the mountains. Here, ox-carts were dangerous and had to be abandoned. It was with difficulty that the oxen could be driven along the narrow ledges.

Jasper House, Whitefish Lake, the ruins of Henry House, they saw from the height of the pass. One foaming stream they forded eight times in three hours, driven from side to side by precipice and windfall; and in places they could advance only by ascending the stream bed. This was risky work on a fractious pony, and some of the riders preferred wading to riding. At noon on the 22nd of August the riders crossed a small stream and set up their tents on the border of a sedgy lake. Then somebody noticed that the lake emptied west, not east; and a wild halloo split the welkin. They had crossed the Divide. They were on the headwaters of the Fraser, where a man could stand astride the stream; and the Fraser led to the Cariboo gold-diggings. They still had four hundred miles to travel. Their boots were in shreds and their clothes in tatters; but what were four hundred miles to men who had tramped almost three thousand?

But their progress had been so slow that the provisions were running short. The first snow of the mountains falls in September, and it was already near the end of August. There was not a moment to lose in resting. What had been a lure of hope now became a goad of desperation. So it is with all life's highest emprises. We plunge in led by hope. We plunge on spurred by fate. When the reward is won, only God and our own souls know that, even if we would, we could not have done otherwise than go on.

Those travellers who had insisted on bringing oxen had now to kill them for meat. Chipmunks were shot for food. So were many worn-out horses. Hides were used to resole boots and make mitts. Not far from Moose Lake the last bag of pemmican was eaten. Perhaps it was a good thing at this time that the band of Overlanders began to spread out and scatter along the trail; for hungry men in large groups are a tragic danger to themselves. Those of the advance-party were now some ten days ahead of their companions in the rear. Mrs MacNaughton, whose husband was with the rear party, of which we shall hear more anon, relates the story of a young fellow so ravenous that he fried the deer-thong he had bought for a tump-line back at one of the company's forts. Fortunately, somewhere west of Moose Lake, the travellers came on a band of Shuswap Indians who traded for matches and powder enough salmon and cranberry cakes to stave off actual famine.

Trees with chipped bark pointed the way down the Fraser. For three days the party followed the little stream that had come out of the lake hardly wider than the span of a man's stride. With each mile its waters swelled and grew wilder. On the third day windfall and precipice drove the riders back from the river bed into the heavy hemlock forest, where festoons of Spanish moss overhead almost shut out the light of the sun and all sense of direction. And when they came back to the bank of the stream they saw a wild cataract cutting its way through a dark canyon. There was no mistake. This was the Fraser, and it was living up to its reputation.

And yet the Overlanders were sorely puzzled. There were no more blazes on the trees to point the way; and, if this was the Fraser, it seemed to flow almost due north. Where was Cariboo? Mr McMicking, who was acting as captain, tried to find out from the Indians. They made him a drawing showing that if he crossed another watershed he would come on a white man's wide pack-road. That must lead to Cariboo; but the snow lay already a foot deep on this road; and unless the Overlanders hastened they would be snow-bound for the winter. On the other hand, if the white men continued to follow the wild river canyon north, it would bring them to Fort George on the main Fraser in ten days. There was no time to waste on chance travelling. The Overlanders knew that somewhere south from Moose Lake must lie the headwaters of the Thompson, which would bring them to Kamloops. Was that what the Indians meant by their drawings of a white man's road? If that were true, between Moose Lake and the Thompson must lie the land of their desire,

Cariboo; but to cross another unknown divide in winter seemed risky. To follow the bend of the Fraser north might be the long way round, but it was sure.

It was decided to let the party separate. Let those with provisions still remaining try to push overland to Cariboo. If they failed to find it, they could build cabins and winter on their pack animals. Twenty men joined this group. The rest decided to stick to the river. Behind were straggling a score more of the travellers, who were left to follow as they could. Mrs Shubert with her children joined the band going overland to find the Thompson.

The Indians traded canoes for horses and showed the Overlanders how to put rafts together to run the Fraser. Axes had been worn almost to the haft. Cutting the huge trees and splitting them into suitable timbers was slow work. It was September before the rafts were ready to be launched. There were four. Each had a heavy railing round it like that of a ferry, with some flat stones on which fires could be lighted to cook meals without pausing to land. When we recall the experiences of Mackenzie and Fraser on this river, it seems almost incredible that these landsmen made the descent on rafts with their few remaining ponies and oxen tied to the railings; yet so they did. If we imagine rafts, with horses and oxen tied to the railings, trying to run the whirlpool below Niagara, we shall have some conception of what this meant.

The canoes sheered out of the way and the rafts were unmoored. The Scarborough raft, with men from Whitby and Scarborough, near Toronto, swirled out to midstream on the afternoon of the 1st of September. 'Poor, poor white men,' sighed the Indians; 'no more see white men'; but the men in the canoes rapped the gunnels with their paddles and uttered rousing cheers. Then the *Ottawa* and the *Niagara* and the *Huntingdon* rafts slipped out on the current. All went well for four days. Sweeps made of trees with the branch ends turned down and long, slim poles kept the rafts in mid-current. Meals were cooked as the unwieldy craft glided along the river-bank. Two or three men kept guard at night, so that the rafts were delayed for only a few hours during the darkest part of the night. The sun shone hot at midday and there were hard frosts at night; but the rest in this sort of travel was wonderfully refreshing after four months of toil across prairie and mountain. But on the afternoon of the 5th of September the rafts began to bounce and swirl. The banks raced to the rear, and before the crews realized it, a noise as of breaking seas filled the air, and the *Scarborough* was riding her first rapid. Luckily, the water was deep and the rocks well submerged. The *Scarborough* ran the rapid without mishap and the other rafts followed. On the next day, however, the waters 'collected' and began running in leaps and throwing back spume. Some one shouted 'Breakers! head ashore!' and the galloping rafts bumped on the bank of the river. The banks here were steep for portaging; and the Scarborough boys, brought up on the lake-front, east of Toronto, decided, come what might, to run the rapids. They let go the mooring-rope and went churning into a whirlpool of yeasty spray. All hands bent their strength to the poles. The raft dipped out of sight, but was presently seen riding safely and calmly below the rapids.

Those watching the *Scarborough* from the bank breathed freely again and plucked up heart; but the worst was yet ahead. The oily calm below the first rapid dropped into another maelstrom of angry waters. Into this the *Scarborough* was drawn by the terrible undertow. For a moment the watchers on the bank could see nothing but the horns of the bellowing, frightened oxen tied to the railing. Then the raft was mounting the waves again. The seaworthiness of a raft is, of course, well known. It may dip under water, or even split, but it seldom upsets and never swamps or sinks. Before the other rafts ran the rapids, two of them were first lightened of their loads. The men preferred to pack their provisions over the precipices rather than take the risk of losing them in the rapid. Nor was the packing child's play. There was a narrow portage-trail along the ledges of the rocks, and where the slabs of granite had split off Indians had laid rickety poles across. Over these frail bridges the packers, with great difficulty, carried the loads of the two rafts. Fortunately most of them had long since discarded boots for moccasins.

All the rafts came through safely. The canoes were not so fortunate. When the *Scarborough* reached a sand-bar at the foot of the rapids, the men were surprised to find three of their Toronto friends, who had gone ahead in a canoe, now stranded high and dry. The canoe had sidled to the waves, swamped, and sunk with everything the Toronto men owned, including their coats, tents, and boots. For two days they had been awaiting the coming of the rafts. They were almost dead from exposure and hunger.

Nine canoes in all were wrecked at this spot. One split on the reef. Another was caught in the backwater. Others sank in the whirlpool below the rapids. Others went under at the first leap into the cataract. Two of the canoes had foolishly been lashed abreast. They sidled, shipped a billow, and sank. All the men clung to the gunnels; but one who was a powerful swimmer struck out for the shore. The canoes stranded on the shore below and the clinging men saved themselves. When they looked for their friend who had struck out for the shore, he was no longer to be seen. These men were all from Goderich, brought up on the banks of Lake Huron.

A similar fate befell a crew of four men from Toronto. Two of them undertook to portage provisions along the bank of the canyon, while the other two, named Carpenter and Alexander, tried to run the canoe down the rapids. The episode has some interest for students of psychology. Carpenter walked down the bank of the canyon a short distance to reconnoitre the different channels of the rapids. He was seen to take out his note-book and write an entry. He then put the note-book in the inner pocket of his coat, took off the coat, and slung it in a tree on the bank. When he came back to the canoe, he seemed preoccupied. The canoe ripped on a rock in midstream, flattened, and sank. Carpenter went down insensible as though his head had struck and he had been stunned. Alexander was washed ashore. He found himself on the side of the bank opposite the rest of the party. Going below to calmer waters, he swam across. Carpenter's coat hung on the trees. In the pocket was the note-book, in which Alexander read the astounding words: 'Arrived at Grand Canyon.

Ran the canyon and was drowned.' Carpenter left a wife and child in Toronto, for whom, evidently, he had written the message. But if he was of sound mind, desiring to live, and so certain of death that he was able to write his own fate in the past tense, why did he attempt the rapids? His friends had no explanation of the curious incident.

There is another gruesome story of a sand-bar in the very middle of this raging canyon. It will be remembered that some of the Overlanders had straggled far to the rear. Some time before spring a party of them attempted to run this canyon. They were never again seen alive. Some treasure-seekers who came over the trail in spring stranded on this sand-bar. They found the bodies of the missing men. All but one had been torn and partly devoured. It need not be told here that no wild beast could have stemmed the rapids from either side. Unless wolves or cougars had accidentally been washed to the sand-bar, and washed away again, the wild solitude must have witnessed a horror too terrible to be told; for the body of the man who had apparently died last was fully clothed and unmolested. As absolutely nothing more is known of what happened than has been set down here, it seems well that there is no record of the names of these castaways.

CHAPTER SIX
QUESNEL AND KAMLOOPS

The walls of the river lowered and widened, the current slackened, and the surviving canoes and rafts were presently gliding peacefully down a smooth stream. That night the Overlanders slept dead with weariness; but a fearful depression rested on the company. Gold had begun to collect its toll, and the price appalled every soul. Who would be the next? How soon would the unknown river turn west and south? Where was Fort George? What perils yet lay between the fort and the gold camp?

As the heavy mists lifted at daybreak, the travellers observed that the river was narrowing again and that the wooded banks had begun to fly past very swiftly. There was no mistaking the signs. They were approaching more rapids. But the trick of guiding the craft down rapids had now been learned; so the flotilla rode the furious waters unharmed for fifteen miles.

It was almost dark when canoes and rafts swung round a curve in the river and saw a flag waving above the little walled fur-post of Fort George. The tired wanderers were welcomed in by clerks too amazed to speak, while a howling chorus of husky-dogs set up their serenade. A young Englishman, who had joined the Overlanders at St Paul, died from the effects of exposure a few minutes after being carried into the fort. Next morning the body was rolled in blankets, placed in a canoe, and buried under a rude wooden cross, with stones piled above the grave to prevent the ravaging of huskies and wolves.

The chief factor was away, but the young clerks in charge sent Indians along to pilot the Overlanders through the rapids below Fort George, known as the most dangerous on the Fraser. These rapids, it will be recalled, had wrecked Alexander Mackenzie and had almost cost Simon Fraser his life. But the treasure-seekers did not have to go as far south as Alexandria, where Mackenzie had turned back. With guides who knew the waters, they ran the rapids below Fort George safely, and moored at Quesnel, the entrance to Cariboo, on the 11th of September—four months after they had left Canada.

Quesnel was at this time a rude settlement of perhaps a dozen log shacks—chiefly bunkhouses and provision-stores. North of Yale the Cariboo Road had not yet been opened, and all provisions had been brought in from the lower Fraser by pack-horse and dog-train at enormous cost and risk. Food sold at extortionate prices. A meal cost two dollars and fifty cents, for beans, bacon, and coffee. Salmon, of course, was cheap. Fortunately, there was little whisky; so, though tattered miners were everywhere in the woods, order was maintained without vigilance committees. On one spectacle the far-travelled ragged Overlanders feasted their tired eyes. They saw miners everywhere along the banks of creeks washing gold. But there were more gold-seekers than claims, and those without claims were full of complaints and fears for the winter. They declared the country was over-rated and a humbug. The question was how 'to get out' to Victoria. Overlanders, who had tramped across the breadth of a continent, did not relish the prospect, as one Yankee miner described it, of 'hoofing it five hundred miles farther.' Some of the disappointed Overlanders floated on down to Alexandria, where they sold their rafts and took jobs on the government road which was being constructed along the canyon. This ensured them safety from starvation for the winter at least.

Other Overlanders followed these first pioneers 'the plains across.' And we have seen that some of those who had crossed the prairie with the first party had fallen behind. These stragglers did not reach Yellowhead Pass till the first week of September. They were entirely out of food; but they had matches, and each box of fifty bought a huge salmon from the Shuswaps.

Some of the men pushed ahead, built a raft, and launched it on the Fraser. The raft ripped on a rock in midstream and stuck there at an angle of forty-five degrees. Money, tools, food, and clothing slithered into the tow of the rapids, while the men clung in desperation to the upper railing of the wreck. One man let go and dropped into the water. Swimming and drifting and rolling over and over, he gained the shore, and hurried back to the pass with word of the accident. Friends, accompanied by Indians, came in canoes to the rescue, and, by means of ropes, every man was brought off the wrecked raft alive.

But the party now stood in a more desperate predicament than ever, for lack of food and clothing. The Shuswaps saved the whites from starvation. They took the white men to a pool in the Fraser, where salmon, exhausted from the long run up the river, could be speared or clubbed by the boat-load. And while some of the men chopped down trees to build dugout canoes, others speared, cleaned, and dried the salmon. Night and day they worked, and forgot sleep in their desperate haste. At length they launched their craft on the Fraser. On the way down the dangerous canyon they saw the wrecked canoes of those who had gone before. The tenth day after leaving Yellowhead Pass they reached Fort George. Their story has been told by Mrs MacNaughton, whose husband was of the party. They arrived at Fort George mostly barefoot, coatless, and trousers and shirts in tatters. Their hair and beards were long and unkempt. It is supposed that they must have lost the salmon in some of the rapids, or else the supply was insufficient; for they were so weak from hunger that they had to be carried into the fort. They arrived at Quesnel a month after the first Overlanders, when the snow was too deep in the mountains for prospecting or mining. The majority of this party also took work on the government road.

Meanwhile, how had fared that band of the Overlanders who had gone over the hills south from the pass in search of the upper branches of the Thompson? A Shuswap accompanied them as guide, and for a few days there was a well-defined game-trail. Then the trail meandered off into a dense forest of hemlock and windfall, which had to be cut almost every mile of the way. They did not average six miles a day; but they finally came to the steep bank of a wild river flowing south which they judged must be a branch of the Thompson. The mountains were so steep that it was impossible to proceed farther with horses and oxen; so they abandoned these in the woods, and cut trees for rafts. For seven days they ran rapid after rapid. One of the rafts stranded on a rock and remained for two days before companions came to the rescue. At another point a canoe was smashed in midstream. The crew struggled to a slippery rock and hung to the ledge. A man named Strachan attempted to swim ashore to signal distress to those above. They saw him ride the waves. Then a roll of angry waters swept over him and he passed out of sight. His companions clung to the rock till another canoe came shooting down-stream, when lines were hoisted to the castaways, and they were hauled ashore.

Where the Clearwater comes into the Thompson they found the fur-trader's horse-trail and tramped the remaining hundred miles overland south to Kamloops. On the last lap of their terrible march all were so exhausted they could scarcely drag themselves forward. Some would lie down and sleep, then creep on a few miles. About twenty miles from the mouth of the Thompson they came to a field of potatoes planted by some rancher of Kamloops. The starving Overlanders could scarcely credit their eyes. No one occupied the windowless log cabin; but there was the potato patch—an oasis of food in a desert of starvation. They paused long enough at the cabin to boil a great kettleful and to feast ravenously. This gave them strength to tramp on to Kamloops. We saw that the Irish mother, Mrs Shubert, with her two children, accompanied this party. The day after reaching Kamloops she gave birth to a child.

Did the Overlanders find the gold which each man's rainbow hopes had dreamed? They had followed the rainbow over the ends of earth. Was the pot of gold at the end of the rainbow? You will find an occasional Overlander passing the sunset of his days in quiet retreat at Yale or Hope or Quesnel or Barkerville. He does not wear evidence of great earthly possessions, though he may refer wistfully to the golden age of those long-past adventurous days. The leaders who survived became honoured citizens of British Columbia. Few came back to the East. They passed their lives in the wild, free, new land that had given them such harsh experiences.

CHAPTER SEVEN

LIFE AT THE MINES

Fortunately, in that winter of '62–'63, there was a great deal of work to be done in the mining country, and men were in high demand. The ordinary wage was ten dollars a day, and men who could be trusted, and who were brave enough to pack the gold out to the coast, received twenty and even as high as fifty dollars a day. There is a letter, written by Sir Matthew Begbie, describing how the mountain trails were infested that winter by desperadoes lying in wait for the miners who came staggering over the trail literally weighted down with gold. The miners found what the great banks have always found, that the presence of unused gold is a nuisance and a curse. They had to lug the gold in leather sacks with them to their work, and back with them to their shacks, and they always carried firearms ready for use. There was very little shooting at the mines, but if a bad man 'turned up missing,' no one asked whether he had 'hoofed' it down the trail, or whether he hung as a sign of warning from a pole set horizontally at a proper height between two trees. In a mining camp there is no mercy for the crook. If the trail could have told tales, there would have been many a story of dead men washed up on the bars, of sneak-thieves given thirty-nine lashes and like the scapegoat turned out into the mountain wilds—a rough-and-ready justice administered without judge or jury.

But a woman was as safe on the trail as in her own home—a thing that civilization never understands about a wild mining camp. Mrs Cameron, wife of the famous Cariboo Cameron, lived with her husband on his claim till she died, and many other women lived in the camps with their husbands. When the road opened, there was a rush of hurdy-gurdy girls for dance-halls; but that did not modify the rough chivalry of an unwritten law. These hurdy-gurdy girls, who tiptoed to the concertina, the fiddle, and the hand-organ, were German; and if we may believe the poet of Cariboo, they were something like the Glasgow girls described by Wolfe as 'cold to everything but a bagpipe—I wrong them—there is not one that does not melt away at the sound of money.' Sings the poet of Cariboo:

> They danced a' nicht in dresses licht
> Fra' late until the early, O!
> But O, their hearts were hard as flint,
> Which vexed the laddies sairly, O!
>
> The dollar was their only love,
> And that they loved fu' dearly, O!
> They dinna care a flea for men,
> Let them court hooe'er sincerely, O!

Cariboo was what the miners call a 'he-camp.' Not unnaturally, the 'she-camps' heard 'the call from Macedonia.' The bishop of Oxford, the bishop of London, the lord mayor of London, and a colonial society in England gathered up some industrious young women as suitable wives for the British Columbia miners. Alack the day, there was no poet to send letters to the outside world on this handling of Cupid's bow and arrow! The comedy was pushed in the most business-like fashion. Threescore young girls came out under the auspices of the society and the Church, carefully shepherded by a clergyman and a stern matron. They reached Victoria in September of '62 and were housed in the barracks. Miners camped on every inch of ground from which the barracks could be watched; and when the girls passed to and from their temporary lodging, their progress was like a royal procession through a silent, gaping, but most respectful lane of whiskered faces. A man looking anything but respect would have been knocked down on the spot. We laugh now! Victoria did not laugh then. It was all taken very seriously. On the instant, every girl was offered some kind of situation, which she voluntarily and almost immediately exchanged for matrimony. In all, some ninety girls came out under these auspices in '62–'63. The respectable girls fitted in where they belonged. The disreputable also found their own places. And the mining camp began to take on an appearance of domesticity and home.

Matthew Begbie, later, like Douglas, given a title for his services to the Empire, had, as we have seen, first come out under direct appointment by the crown; and when parliamentary government was organized in British Columbia his position was confirmed as chief justice. He had less regard for red tape than most chief justices. Like Douglas, he first maintained law and order and then looked up to see if he had any authority for it. No man ever did more for a mining camp than Sir Matthew Begbie. He stood for the rights of the poorest miner. In private life he was fond of music, art, and literature; but in public life he was autocratic as a czar and sternly righteous as a prophet. He was a vigilance committee in himself through sheer force of personality. Crime did not flourish where Begbie went. Chinaman or Indian could be as sure of justice as the richest miner in Cariboo. From hating and fearing him, the camp came almost to worship him.

Many are the stories of his circuits. Once a jury persisted in bringing in a verdict of manslaughter in place of murder.

'Prisoner,' thundered Begbie, 'it is not a pleasant duty to me to sentence you *only* to prison for life. You deserve to be hanged. Had the jury performed their duty, I might have the painful satisfaction of condemning you to death. You, gentlemen of the jury, permit me to say that it would give me great pleasure to sentence you to be hanged each and every one of you, for bringing in a murderer guilty only of manslaughter.'

On another occasion, when an American had 'accidentally' shot an Indian, the coroner rendered a verdict 'worried to death by a dog.' Begbie ordered another inquest. This time the coroner returned a finding that the Indian 'had been killed by falling over a cliff.' Begbie on his own authority ordered the American seized and taken down to Victoria. On his way down the prisoner escaped from the constable. This type of hair-trigger gunmen at once fled the country when Begbie came.

Mr Alexander, one of the Overlanders of '62, tells how 'Begbie's decisions may not have been good law, but they were first-class justice.' His 'doctrine was that if a man were killed, some one had to be hanged for it; and the effect was salutary.' A man had been sandbagged in a Victoria saloon and thrown out to die. His companion in the saloon was arrested and tried. The circumstantial evidence was strong, and the judge so charged the jury. But the jury acquitted the prisoner. Dead silence fell in the court-room. The prisoner's counsel arose and requested the discharge of the man. Begbie whirled: 'Prisoner at the bar, the jury have said you are not guilty. You can go, and I devoutly hope the next man you sandbag will be one of the jury.' On another occasion a man was found stabbed on the Cariboo Road. The man with whom the dead miner had been quarrelling was arrested, tried, and, in spite of strong evidence against him, acquitted. Begbie adjourned the court with the pious wish that the murderer should go out and cut the throats of the jury.

But, in spite of his harsh manner towards the wrong-doer, 'the old man,' as the miners affectionately called him, kept law and order. In the early days gold commissioners not only settled all mining disputes, but acted as judge and jury. Against any decision of the gold commissioners Begbie was the sole appeal, and in all the long years of his administration no decision of his was ever challenged.

The effect of sudden wealth on some of the hungry, ragged horde who infested Cariboo was of a sort to discount fiction. One man took out forty thousand dollars in gold nuggets. A lunatic escaped from a madhouse could not have been more foolish. He came to the best saloon of Barkerville. He called in guests from the highways and byways and treated them to champagne which cost thirty dollars and fifty dollars a bottle. When the rabble could drink no more champagne, he ordered every glass filled and placed on the bar. With one magnificent drunken gesture of vainglory he swept the glasses in a clattering crash to the floor. There was still a basket of champagne left. He danced the hurdy-gurdy on that basket till he cut his feet. The champagne was all gone, but he still had some gold nuggets. There was a mirror in the bar-room valued at hundreds of dollars. The miner stood and proudly surveyed his own figure in the glass. Had he not won his dearest desire and conquered all things in conquering fortune? He gathered his last nuggets and hurled them in handfuls at the mirror, shattering it in countless pieces. Then he went out in the night to sleep under the stars, penniless. He settled down to work for the rest of his life in other men's mines.

The staid Overlanders, who had risked their lives to reach this wild land of desire, who had come from such church-going hamlets as Whitby, such Scottish-Presbyterian centres as Toronto and Montreal, hardly knew whether they were dreaming or living in a country of crazy pixies who delved in mud and water all day and weltered in champagne all night. The Cariboo poet sang their sentiments in these words:

> I ken a body made a strike.
>   He looked a little lord.
> He had a clan o' followers
>   Amang a needy horde.

> Whane'er he'd enter a saloon,
>   You'd see the barkeep smile?
> His lordship's humble servant he
>   Wi'out a thought o' guile!

A twalmonth passed an' a' is gane,
    Baith freends and brandy bottle!
An' noo the puir soul's left alane
    Wi' nocht to weet his throttle!

In Barkerville, which became the centre of Cariboo, saloons and dance-halls grew up overnight. Pianos were packed in on mules at a rate of a dollar a pound from Quesnel. Champagne in pint bottles sold at two ounces of gold. Potatoes retailed at ninety dollars a hundredweight. Nails were cheap at a dollar a pound. Milk was retailed frozen at a dollar a pound. Boots still cost fifty dollars. Such luxuries as mirrors and stoves cost as high as seven hundred dollars each. The hurdy-gurdy girls with true German thrift charged ten dollars or more a dance—not the stately waltz, but a wild fling to shake the rafters and tire out the stoutest miners.

A newspaper was published in Barkerville. And it was in it that James Anderson of Scotland first issued *Jeames's Letters to Sawney*.

Your letter cam' by the express,
Eight shillin's carriage, naethin' less!
You maybe like to ken what pay
Miners get here for ilka day?
Jus' twa poond sterling', sure as death?
It should be four, between us baith?
For gin ye coont the cost o' livin',
There's naethin' left to gang an' come on.
Sawney, had ye yer taters here
And neeps and carrots—dinna speer
What price; though I might tell ye weel,
Ye'd ainly think me a leein' chiel.

The first twa years I spent out here
Werena sae ill ava';
But hoo I've lived syne; my freend,
There's little need to blaw.

Like fitba' knockit back and fore,
That's lang in reachin' goal,
Or feather blown by ilka wind
That whistles 'tween each pole?
E'en sae my mining life has been
For mony a weary day.

Later, when the dance-hall became the theatre of Barkerville,
James Anderson used to sing his rhymes to the stentorious shouting
and loud stamping of the shirt-sleeved audience.

He thinks his pile is made,
An' he's goin' hame this fall,
To join his dear auld mither,
His faither, freends, and all.
His heart e'en jumps wi' joy
At the thocht o' bein' there,
An' mony a happy minute
He's biggin' castles in the air!

But hopes that promised high
In the springtime o' the year,
Like leaves o' autumn fa'
When the frost o' winter's near.
Sae his biggin' tumbles doon,
Wi' ilka blast o' care,
Till there's no stane astandin'
O' his castles in the air.

The Cariboo Road

CHAPTER EIGHT

THE CARIBOO ROAD

When the railway first went through the Fraser Canyon, passengers looking out of the windows anywhere from Yale to Ashcroft were amazed to see something like a Jacob's ladder up and down the mountains, appearing in places to hang almost in mid-air. Between Yale and Lytton it hugged the mountain-side on what looked like a shelf of rock directly above the wildest water of the canyon. Cribwork of huge trees, resembling in the distance the woven pattern of a willow basket, projected out over the ledges like a bird's nest hung from some mountain eyrie. The traveller almost expected to see the thing sway and swing to the wind. Then the train would sweep through a tunnel, or swing round a sharp bend, and far up among the summits might be seen a mule-team, or a string of pack-horses winding round the shoulders of the rock. It seemed impossible that any man-made highway could climb such perpendicular walls and drop down precipitous cliffs and follow a trail apparently secure only for a mountain goat. The first impression was that the thing must be an old Indian war-path, along which no enemy could pursue. But when the train paused at a water tank, and the traveller made inquiry, he was told that this was nothing less than the famous Cariboo Road, one of the wonders of the world.

As long as the discovery of gold was confined to the Fraser river-bars, the important matter of transportation gave the government no difficulty. Hudson's Bay steamers crossed from Victoria to Langley on the Fraser, which was a large fort and well equipped as a base of supplies for the workers in the wilderness. Stern-wheelers, canoes, and miscellaneous craft could, with care, creep up from Langley to Hope and Yale; and the fares charged afforded a good revenue to the Hudson's Bay Company. Even when prospectors struck above Yale, on up to Harrison Lake and across to Lillooet, or from the Okanagan to the Thompson, the difficulties of transportation were soon surmounted. A road was shortly opened from Harrison Lake to Lillooet, built by the miners themselves, under the direction of the Royal Engineers; and, as to the Thompson, there was the well-worn trail of the fur-traders, who had been going overland to Kamloops for fifty years.

It was when gold was discovered higher up on the Fraser and in Cariboo, after the colony of British Columbia had taken its place on the political map, that Governor Douglas was put to the task of building a great road. Henceforth, for a few years at least, the miners would be the backbone, if not the whole body, of the new colony. How could the administration be carried on if the government had no road into the mining region?

And so the governor of British Columbia entered on the boldest undertaking in roadbuilding ever launched by any community of twenty thousand people. The Cariboo Road became to British Columbia what the Appian Way was to Rome. It was eighteen feet wide and over four hundred and eighty miles long. It was one of the finest roads ever built in the world. Yet it cost the country only two thousand dollars a mile, as against the forty thousand dollars a mile which the two transcontinental railways spent later on their roadbeds along the canyon. It was Sir James Douglas's greatest monument.

Five hundred volunteer mine-workers built the road from Harrison Lake to Lillooet in 1858 at the rate of ten miles a day; and when the road was opened in September, packers' charges fell from a dollar to forty-eight cents and finally to eighteen cents a pound. But presently the trend of travel drew away from Harrison Lake to the line of the Fraser. At first there was nothing but a mule-trail hacked out of the rock from Yale to Spuzzum; but miners went voluntarily to work and widened the bridle-path above the shelving waters. From Spuzzum to Lytton the river ledges seemed almost impassable for pack animals; yet a cable ferry was rigged up at Spuzzum and mules were sent over the ledges to draw it up the river. When the water rose so high that the lower ledges were unsafe, the packers ascended the mountains eight hundred feet above the roaring canyon. Where cliffs broke off, they sent the animals across an Indian bridge. The marvel is not that many a poor beast fell headlong eight hundred feet down the precipice. The marvel is that any pack animal could cross such a trail at all. 'A traveller must trust his hands as much as his feet,' wrote Begbie, after his first experience of this trail.

But by 1862 cutting and blasting and bridge-building had begun under the direction of the Royal Engineers; and before 1865 the great road was completed into the heart of the mining country at Barkerville. Henceforth passengers went in by stage-coach drawn by six horses. Road-houses along the way provided relays of fresh horses. Freight went in by bull-team, but pack-horses and mules were still used to carry miners' provisions to the camps in the hills which lay off the main road. It was while the road was still building that an enterprising packer brought twenty-one camels on the trail. They were not a success and caused countless stampedes. Horses and mules took fright at the slightest whiff of them. The camels themselves could stand neither the climate nor the hard rock road. They were turned adrift on the Thompson river, where the last of them died in 1905.

There was something highly romantic in the stage-coach travel of this halcyon era. The driver was always a crack whip, a man who called himself an 'old-timer,' though often his years numbered fewer than twenty. Most of the drivers, however, knew the trail from having packed in on shanks's mare and camped under the stars. At the log taverns known as road-houses travellers could sleep for the night and obtain meals.

Indian graves at Lytton, BC
FROM A PHOTOGRAPH

On the down trip bags were piled on the roof with a couple of frontiersmen armed with rifles to guard them. Many were the devices of a returning miner for concealing the gold which he had won. A fat hurdy-gurdy girl—or sometimes a squaw—would climb to a place in the stage. And when the stage, with a crack of the whip and a prance of the six horses, came rattling across the bridge and rolling into Yale, the fat girl would be the first to deposit her ample person at the bank or the express office, whence gold could safely be sent on down to Victoria. And when she emerged half an hour later she would have thinned perceptibly. Then the rough miner, who had not addressed a word to her on the way down, for fear of a confidence man aboard, would present 'Susy' with a handsome reward in the form of a gaudy dress or a year's provisions.

Start from a road-house was made at dawn, when the clouds still hung heavy on the mountains and the peaks were all reflected in the glacial waters. The passengers tumbled dishevelled from log-walled rooms where the beds were bench berths, and ate breakfast in a dining-hall where the seats were hewn logs. The fare consisted of ham fried in slabs, eggs ancient and transformed to leather in lard, slapjacks, known as 'Rocky Mountain dead shot,' in maple syrup that never saw a maple tree and was black as a pot, and potatoes in soggy pyramids. Yet so keen was the mountain air, so stimulating the ozone of the resinous hemlock forests, that the most fastidious traveller felt he had fared sumptuously, and gaily paid the two-fifty for the meal. Perhaps there was time to wash in the common tin basin at the door, where the towel always bore evidence of patronage; perhaps not; anyhow, no matter. Washing was only a trivial incident of mountain travel in those days.

The passenger jumped for a place in the coach; the long whip cracked. The horses sprang forward; and away the stage rattled round curves where a hind wheel would try to go over the edge—only the driver didn't let it; down embankments where any normal wagon would have upset, but this one didn't; up sharp grades where no horses ought to be driven at a trot, but where the six persisted in going at a gallop! The passenger didn't mind the jolting that almost dislocated his spine. He didn't mind the negro who sat on one side of him or the fat squaw who sat on the other. He was thankful not to be held up by highwaymen, or dumped into the wild cataract of waters below. Outside was a changing panorama of mountain and canyon, with a world of forests and lakes. Inside was a drama of human nature to outdo any curtain-raiser he had ever witnessed—a baronet who had lost in the game and was going home penniless, perhaps earning his way by helping with the horses; an outworn actress who had been trying her luck at the dance-halls; a gambler pretending that he was a millionaire; a saloon-keeper with a few thousands in his pockets and a diamond in his shirt the size of a pebble; a tenderfoot rigged out as a veteran, with buckskin coat, a belt full of artillery, fearfully and wonderfully made new high-boots, and a devil-may-care air that deceived no one but himself; a few Shuswaps and Siwashes, fat, ill-smelling, insolent, and plainly highly amused in their beady, watchful, black, ferret eyes at the mad ways of this white race; a still more ill-smelling Chinaman; and a taciturn, grizzled, ragged fellow, paying no attention to the fat squaw, keeping his observations and his thoughts inside his high-boots, but likely as not to turn out the man who would conduct the squaw to the bank or the express office at Yale.

If one could get a seat outside with the guards and the driver—one who knew how to unlock the lore of these sons of the hills—he was lucky; for he would learn who made his strike there, who was murdered at another place, how the sneak-thief trailed the tenderfoot somewhere else—all of it romance, much of it fiction, much of it fact, but no fiction half so marvellous as the fact.

Bull-teams of twenty yokes, long lines of pack-horses led by a bell-mare, mule-teams with a tinkling of bells and singing of the drivers, met the stage and passed with happy salute. At nightfall the camp-fires of foot travellers could be seen down at the water's edge. And there was always danger enough to add zest to the journey. Wherever there are hordes of hungry, adventurous men, there will be desperadoes. In spite of Begbie's justice, robberies occurred on the road and not a few murders. The time going in and out varied; but the journey could be made in five days and was often made in four.

The building of the Cariboo Road had an important influence on the camp that its builders could not foresee. The unknown El Dorado is always invested with a fabulous glamour that draws to ruin the reckless and the unfit. Before the road was built adventurers had arrived in Cariboo expecting to pick up pails of nuggets at the bottom of a rainbow. Their disillusionment came; but there was an easy way back to the world. They did not stay to breed crime and lawlessness in the camp. 'The walking'—as Begbie expressed it—'was all down hill and the road was good, especially for thugs.' While there were ten thousand men in Cariboo in the winter of '62 and perhaps twenty thousand in the winter of '63, there were less than five thousand in '71.

This does not mean that the camp had collapsed. It had simply changed from a poor man's camp to a camp for a capitalist or a company. It will be remembered that the miners first found the gold in flakes, then farther up in nuggets, then that the nuggets had to be pursued to pay-dirt beneath gravel and clay. This meant shafts, tunnels, hydraulic machinery, stamp-mills. Later, when the pay-dirt showed signs of merging into quartz, there passed away for ever the day of the penniless prospector seeking the golden fleece of the hills as his predecessor, the trapper, had sought the pelt of the little beaver.

All unwittingly, the miner, as well as the trapper, was an instrument in the hands of destiny, an instrument for shaping empire; for it was the inrush of miners which gave birth to the colony of British Columbia. Federation with the Canadian Dominion followed in 1871; the railway and the settler came; and the man with the pick and his eyes on the 'float' gave place to the man with the plough.

ENDNOTES

CHAPTER ONE

[1] The same, of course, may be done to-day, with a like result, at many places along the Fraser and even on the Saskatchewan.

[2] This was the first Legislative Assembly to meet west of Upper Canada in what is now the Canadian Dominion. It consisted of seven members, as follows: J.D. Pemberton, James Yates, E.E. Langford, J.S. Helmcken, Thomas J. Skinner, John Muir, and J.F. Kennedy. Langford, however, retired almost immediately after the election and J.W. McKay was elected in his stead. The portraits of five of the members are preserved in the group which appears as the frontispiece to this volume. The photograph was probably taken at a later period; at any rate, two of the members, Muir and Kennedy, are missing.

CHAPTER THREE

[1] See *Pioneers of the Pacific Coast* in this Series.

CHAPTER FOUR

[1] See the map in *The Adventurers of England on Hudson Bay* in this Series.

[2] Perhaps the distinction should be made here between the muskeg and the slough. The slough was simply any depression in the ground filled with mud and water. The muskeg was permanent wet ground resting on soft mud, covered over on the top with most deceiving soft green moss which looked solid, but which quaked to every step and gave to the slightest weight. Many muskegs west of Edmonton have been formed by beavers damming the natural drainage of a small river for so many centuries that the silt and humus washed down from the mountains have formed a surface of deep black muck.

BIBLIOGRAPHICAL NOTE

The episode of Cariboo is so recent that the bibliography on it is not very complete. *British Columbia*, by Judge Howay and E.O.S. Scholefield, provincial librarian, is the last and most accurate word on the history of that province, though one could wish that the authors had given more human-document records in the biographical section. In a very few years there will be no old-timers of the trail left; and, after all, it is the human document that gives colour and life to history. It was my privilege to know some of the Overlanders intimately. One of the companies who rafted down the Fraser came from the county where I was born; and though they preceded my day, their terrible experiences were a household word. With others I have poled the Fraser on those very tempestuous waters that took such toll of life in '62. Others have been my hosts. I have gone up and down the Arrow Lakes in a steamer as a guest of the man who came through the worst experiences of the Overlanders. Chance conversations are shifty guides on dates and place-names. For these, regarding the Overlanders, I have relied on Mrs MacNaughton's *Cariboo*.

Gosnell's *British Columbia Year Book* and Hubert Howe Bancroft's *British Columbia* are very full on this era. Walter Moberly's pamphlets on the building of the trail and Mr Alexander's casual addresses are excellent. Old files of the Kamloops *Sentinel* and the Victoria *Colonist* are full of scattered data. Anderson's *Hand Book of 1858*, Begbie's Report to the London Geographical Society, 1861; Begg's *British Columbia*; *Fraser's Journal*; Mayne's *British Columbia*, 1862; Milton and Cheadle's *North West Passage*, 1865; Palliser's *Report*, 1859; Waddington's *Fraser River Mines*—all afford sidelights on this adventurous era. On the prospector's daily life there is no book. That must be learned from him on the trail; and on many camp trips in the Rockies, with prospectors for guides, I have picked up such facts as I could.

INDEX

THE CLASSICS WEST COLLECTION
Capturing the spirit of the Canadian West

 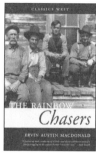  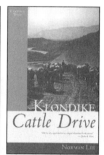

   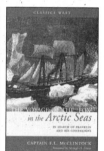

The Ranch on the Cariboo
Alan Fry
978-1-926741-00-0 · 5.5 x 8.5 · 288 pages · $19.95 pb

The Rainbow Chasers
Ervin Austin MacDonald
978-1-894898-30-0 · 5.5 x 8.5 · 288 pages · $19.95 pb

Three Against the Wilderness
Eric Collier
978-1-894898-54-6 · 5.5 x 8.5 · 320 pages · $19.95 pb

Klondike Cattle Drive
Norman Lee
978-1-894898-14-0 · 5.5 x 8.5 · 96 pages · $12.95 pb

Pioneers of the Pacific Coast
Agnes C. Laut
978-1-926971-00-1 · 5.5 x 8.5 · 144 pages · $14.95 pb

The Arctic Journals of John Rae
978-1-927129-74-6 · 5.5 x 8.5 · 320 pages · $19.95 pb

The Discovery of a Northwest Passage
Sir Robert McClure
978-1-77151-009-7 · 5.5 x 8.5 · 288 pages · $19.95 pb

The Voyage of the 'Fox' in the Arctic Seas
Captain F.L. McClintock
978-1-927129-19-7 · 5.5 x 8.5 · 320 pages · $19.95 pb

AGNES C. LAUT was born in Huron County, Ontario, in 1871. She was raised and educated in Winnipeg, and worked briefly as a teacher prior to a bout of tuberculosis that changed the course of her life. She became a reporter and editorial writer for the *Manitoba Free Press* in the 1890s, then a wide-ranging travel writer. Her books include *Pioneers of the Pacific Coast, Lords of the North, The Story of the Trapper, Pathfinders of the West, Vikings of the Pacific, Canada at the Crossroads*, and *The Romance of the Rails*. She died in 1936.

DIANA FRENCH moved to the Cariboo Chilcotin region in 1951 to teach in a one-room school. She married Bob French, the son of a pioneer family, and they settled in Williams Lake in 1970. Along with raising five sons, Diana worked for the *Williams Lake Tribune* as a reporter, and now writes a weekly column for the newspaper. She is involved in various community activities and has been the volunteer curator of the Museum of the Cariboo Chilcotin for more than twenty years. She has also written three books: *The Road Runs West: A Century Along the Bella Coola/Chilcotin Road* (Harbour Publishing); *Ranchland: British Columbia's Cattle Country*, co-authored with Rick Blacklaws (Harbour Publishing); and *Women of Brave Mettle: More Stories of the Cariboo Chilcotin* (Caitlin Press).